Perioperative Neuroscience

Perioperative Neuroscience

Translational research

Edited by

Hemanshu Prabhakar
Department of Neuroanaesthesiology and Critical Care, All India Institute of Medical Sciences, New Delhi, India

Charu Mahajan
Department of Neuroanaesthesiology and Critical Care, All India Institute of Medical Sciences, New Delhi, India

Indu Kapoor
Department of Neuroanaesthesiology and Critical Care, All India Institute of Medical Sciences, New Delhi, India

ELSEVIER

ACADEMIC PRESS
An imprint of Elsevier

Academic Press is an imprint of Elsevier
125 London Wall, London EC2Y 5AS, United Kingdom
525 B Street, Suite 1650, San Diego, CA 92101, United States
50 Hampshire Street, 5th Floor, Cambridge, MA 02139, United States
The Boulevard, Langford Lane, Kidlington, Oxford OX5 1GB, United Kingdom

British Library Cataloguing-in-Publication Data
A catalogue record for this book is available from the British Library

Library of Congress Cataloging-in-Publication Data
A catalog record for this book is available from the Library of Congress

ISBN: 978-0-323-91003-3

For Information on all Academic Press publications
visit our website at https://www.elsevier.com/books-and-journals

Publisher: Nikki P. Levy
Acquisitions Editor: Joslyn Chaiprasert-Paguio
Editorial Project Manager: Susan Ikeda
Production Project Manager: Selvaraj Raviraj
Cover Designer: Christian J. Bilbow

Typeset by MPS Limited, Chennai, India

Dedication

To my teachers, my parents, and my family—Pallavi, Anavi, and Amyra.

–Hemanshu Prabhakar

To my teachers, my parents, and my family—Atul, Namyah, and Nyra.

–Charu Mahajan

To my husband Deepak and my son Ansh, with whom my world is beautiful.

–Indu Kapoor

Contents

List of contributors

Alaa Abd-Elsayed Pain Management Services, UW Health, Madison, WI, United States

Kulikov Alexander Burdenko National Medical Research Center of Neurosurgery, Moscow, Russia

Denise Battaglini Anesthesia and Intensive Care, San Martino Policlinico Hospital, IRCCS for Oncology and Neuroscience, Genoa, Italy; Department of Medicine, University of Barcelona (UB), Barcelona, Spain

Sakhi Bhansali Department of Neurology, Institute of Neurosciences Kolkata, Kolkata, India

Suparna Bharadwaj Department of Neuroanaesthesia and Neurocritical Care, National Institute of Mental Health and Neurosciences, Bengaluru, India

Federico Billotta Anaesthesia and Intensive Care Medicine, Sapienza University of Rome, Rome, Italy

Summit Bloria Department of Anaesthesia and Intensive Care, PGIMER, Chandigarh, India

Ryan R. Budwany Center for Integrative Pain Management, West Virginia University, Morgantown, WV, United States

Marco Cascella Unit of Pain Medicine and Research in AnesthesiaIstituto Nazionale Tumori, IRCCS Fondazione Pascale, Napoli, Italy

Rajeev Chauhan Department of Anaesthesia and Intensive Care, PGIMER, Chandigarh, India

Supriyo Choudhury Department of Neurology, Institute of Neurosciences Kolkata, Kolkata, India

Chandril Chugh MAX Super Speciality Hospital, New Delhi, India

Ahmed Ezzat Department of Surgery and Cancer, Imperial College Healthcare NHS Trust, London, United Kingdom; Department of Surgery and Cancer, Faculty of Medicine, Imperial College London, London, United Kingdom

Bilotta Federico Department of Anesthesiology and Critical Care, Policlicnico Umberto I, "Sapienza" University of Rome, Rome, Italy

Zachary Fleissner Department of Anesthesiology and Perioperative Medicine, Mayo Clinic-Florida, Jacksonville, FL, United States

Benjamin F. Gruenbaum Department of Anesthesiology and Perioperative Medicine, Mayo Clinic-Florida, Jacksonville, FL, United States

Shaun E. Gruenbaum Department of Anesthesiology and Perioperative Medicine, Mayo Clinic-Florida, Jacksonville, FL, United States

Nidhi Gupta Department of Neuroanaesthesia, Indraprastha Apollo Hospitals, New Delhi, India

Indu Kapoor Department of Neuroanesthesiology and Critical Care, All India Institute of Medical Sciences (AIIMS), New Delhi, India

Masahiko Kawaguchi Department of Anesthesiology, Nara Medical University, Kashihara, Japan

Hrishikesh Kumar Department of Neurology, Institute of Neurosciences Kolkata, Kolkata, India

Ankur Luthra Department of Anaesthesia and Intensive Care, PGIMER, Chandigarh, India

Charu Mahajan Department of Neuroanesthesiology and Critical Care, All India Institute of Medical Sciences (AIIMS), New Delhi, India

Russi Mariarita Department of Anesthesiology and Critical Care, Policlicnico Umberto I, "Sapienza" University of Rome, Rome, Italy

Yusuke Naito Department of Anesthesiology, Nara Medical University, Kashihara, Japan

Yeshvant A. Navalgund Center for Integrative Pain Management, West Virginia University, Morgantown, WV, United States

Sangeetha R. Palaniswamy Department of Neuroanaesthesia and Neurocritical Care, National Institute of Mental Health and Neurosciences, Bengaluru, India

Mariangela Panebianco Department of Molecular and Clinical Pharmacology, Institute of Translational Medicine, University of Liverpool, Clinical Sciences Centre for Research and Education, Liverpool, United Kingdom

Neil Patel Department of Anesthesiology and Perioperative Medicine, Mayo Clinic-Florida, Jacksonville, FL, United States

Paolo Pelosi Anesthesia and Intensive Care, San Martino Policlinico Hospital, IRCCS for Oncology and Neuroscience, Genoa, Italy; Department of Surgical Sciences and Integrated Diagnostics, University of Genoa, Genoa, Italy

Hemanshu Prabhakar Department of Neuroanesthesiology and Critical Care, All India Institute of Medical Sciences (AIIMS), New Delhi, India

Zena Radi Sapienza University of Rome, Rome, Italy

Vanitha Rajagopalan Department of Neuroanesthesiology and Critical Care, All India Institute of Medical Sciences (AIIMS), New Delhi, India

Chiara Robba Anesthesia and Intensive Care, San Martino Policlinico Hospital, IRCCS for Oncology and Neuroscience, Genoa, Italy; Department of Surgical Sciences and Integrated Diagnostics, University of Genoa, Genoa, Italy

Francesca Rubulotta Chair of the Department of Critical Care, McGill University, School of Medicine and Health Science, Montreal, QC, Canada

Antonia Schonwald New York Medical College, Valhalla, NY, United States

Sameer Sethi Department of Anaesthesia and Intensive Care, PGIMER, Chandigarh, India

Vasudha Singhal Department of Neuroanesthesiology and Critical Care Medanta-the Medicity, Gurgaon, India

Valerie A. Smart Department of Aviation Medicine, Naval Hospital Jacksonville, Jacksonville, FL, United States

Mona Tiwari Department of Neuro-Radiology, Institute of Neurosciences Kolkata, Kolkata, India

Foreword

Research in medicine has grown enormously over the past 30–40 years and the number of publications has increased tremendously. However, the amount of research that has reached the bedside has not increased proportionately. This phenomenon is most pronounced in neurosciences in which the disease complexity is profound and perplex. Basic science research shows some promising results, but when subjected to clinical trials, it fails to attain the intended goal that is to improve the patients' outcome. This is because the clinical trials are conducted prematurely. That is where the role of translational research lies.

Translational research is the process of applying knowledge from basic research and clinical trials to techniques and tools that address bedside medical needs. The goal of translational research is to move basic science discoveries quickly and efficiently into clinical practice.

There are many areas in neurosciences that require systematic research through a translational pathway. In this respect, the efforts of Dr. Hemanshu Prabhakar and his team are laudable. This book is an effort to identify the areas in which translational research is expected. The editors have chosen international authorities in neurosciences to put together the issues that need to be explored through a translational pathway. In addition, they also have chapters on the meaning of translational research and the guidelines for the conduct of translational research in neurosciences.

Several drugs and physiological techniques have been subjected to experimental and small clinical trials in protecting the ischemic brain from neuronal damage without success. Except for the benefit accrued by correction of abnormal extracerebral physiology, no brain-oriented therapy has been successful. The complex pathophysiology of traumatic brain injury has been the subject of several trials without any major breakthrough. The molecular mechanisms of anesthetic agents have not been understood to date, leave alone their brain-protective potential. Tools to monitor the depth of anesthesia currently available are far from satisfactory and not universally standardized. A serious debate has been going on the neurodevelopmental anomalies in infants exposed to anesthetics. Predictive factors and preventive measures to avoid postoperative cognitive dysfunction in the elderly have not been understood till now. Ideal and practical bedside techniques of intraoperative and critical care monitoring of patients with cerebral pathology have not been expounded till now. Disease-specific special considerations in translational research in stroke, TBI, movement disorders, epilepsy, delayed cerebral ischemia, and pain require an explanation.

All the discussed issues have been carefully addressed in this book. This book will be useful not only to guide future researchers in neurosciences but also clinicians and students to understand the current state of knowledge in the previously mentioned areas.

Dr. G.S. Umamaheswara Rao
Senior Professor, Department of Neuronaesthesia,
National Institute of Mental Health and Neurosciences, Bangalore, India

Acknowledgment

We wish to acknowledge the support from the administration of the All India Institute of Medical Sciences (AIIMS), New Delhi (India), for allowing us to carry out this academic task.

We thank the faculty and staff of the Department of Neuroanaesthesiology and Critical Care at AIIMS, New Delhi, for their support.

Special thanks are due to the team of Elsevier—Melanie Tucker, Susan Ikeda, Joslyn T Chaiprasert-Paguio, R Selvaraj, and all those involved in this project.

Hemanshu Prabhakar
Charu Mahajan
Indu Kapoor

Introduction

1

Indu Kapoor, Charu Mahajan and Hemanshu Prabhakar
Department of Neuroanesthesiology and Critical Care, All India Institute of Medical Sciences (AIIMS), New Delhi, India

Translational research is the process of applying basic biology and clinical trials knowledge to techniques and interventions that address critical needs. This kind of research is basically designed to improve health care outcomes. It consists of a team of experts who focus on translating useful information from laboratories to health care providers, institutes, and hospitals. Translational research can be of three types: T1—developing treatments and interventions; T2—testing the efficacy and effectiveness of these treatments and interventions; and T3—dissemination and implementation research for system-wide change. The term "translational research" was first used in the late 1990s.[1] In the last decades of the 20th century, discovery and spending in the basic medical sciences increased significantly, but the improvement in health care outcomes remained very slow. One of the major reasons was found to be the lag between basic discovery and the appearance of new drugs and treatments. This lag time ranged from 10–20 years and the cost for developing a successful drug was found to be very high.[2] Another important reason for this gap between discovery and clinical application was the decrease in the number of clinician-scientists. In response to that, there has been a drive for more clinician-scientists which should help close this time gap.[3]

Translational research and neurosciences

Translational research in the medical field has become an attractive word. Many research institutes and universities are increasingly developing their translational research credentials. In neurosciences, apart from neurosurgery and neurology, neuroanesthesia has also emerged as a superspeciality discipline for now nearly five decades.[4] There are several areas of interest where basic research needs to be translated to clinical benefits, which include pain modulation, neurotransmission, neurotoxicity, neuro-protection, cognitive awareness, spinal cord and cerebral blood flow, and neurotrauma.[5] In the past, neuroanesthesiologists have conducted basic research in this field, but translating it to clinical settings has not been successful. However, it has been observed that some amount of research work in neuroanesthesia has been translated to actual clinical scenarios.[6–14] The scope of translational research in neuroanesthesia is immense. The *neuro-toxic and neuro-protective effects of anesthetic agents* are the most commonly studied research areas that have been further translated to the clinical setting. Several studies have confirmed the

Perioperative Neuroscience. DOI: https://doi.org/10.1016/B978-0-323-91003-3.00011-8

immediate neurotoxic effect of anesthetic agents in immature rodents; however, these research papers did not study the long-term cognitive consequences in these animals.[6,7] The rodent studies using isoflurane[6] and sevoflurane[7] have confirmed the early neurotoxicity at a cellular level but differed significantly with respect to long-term cognitive outcomes. However, behavioral deficits persisting into adulthood have been found in the clinical research using sevoflurane but not isoflurane. Among intravenous anesthetic agents like thiopental, propofol, ketamine, which act on either N-methyl-D-aspartate (NMDA) or gamma aminobutyric acid (GABA) receptors, studies suggest that when administered either alone or in combination to mice, they can also cause behavioral alterations in adulthood.[8,9] Overall, this evidence suggests that different anesthetic agents with different molecular mechanisms of action can potentially increase cerebral apoptosis in rodents and can lead to long-lasting neuro-behavioral consequences. In an order to cause neurotoxicity, dosage and duration of administration of anesthetic agents should coincide with the brain growth spurt as well as the period of intense neurogenesis and synaptogenesis.[10] These anesthetic agents produce excitotoxicity by acting on either GABA[11] or NMDA receptors. After the withdrawal of the NMDA blockers (ketamine), increased calcium influx could lead to both apoptotic and necrotic cell death.[12]It has been observed that the blockade of NMDA glutamate receptors for even for few hours can trigger widespread apoptotic neuro-degeneration in the developing rat brain.[13]The early research examining the cerebral protective effect of drugs started with anesthetic agents. The first reports to suggest the cerebroprotective effect of anesthetic agents included sodium thiopental. During carotid endarterectomy, it has been observed that sodium thiopental might improve the neurological outcome.[14] Michenfelder et al. examined the patterns of cerebral metabolism in a dog that continuously received thiopental.[15] They found that the electroencephalogram (EEG) became isoelectric following the administration of thiopental, but that further dosages of thiopental thereafter did not result in further suppression of cerebral metabolism. He established the basic concept behind thiopental-induced suppression of cerebral metabolism. Thiopental also causes therapeutic inhibition of global protein synthesis, protects neurons from hypoxic damage by preserving energy balance in oxygen-deprived cells.[16] It has been observed that pharmacological agents that either act directly on $GABA_A$ receptors or modulate $GABA_A$ receptor activity (such as propofol, midazolam) are capable of reducing the severity of brain injury following ischemia in gerbils. It has been reported that at clinically relevant concentrations, propofol inhibits glutamate release by blocking current through sodium channels or by activating $GABA_A$ receptors. Studies have assessed that by intracerebroventricular administration of propofol, 3 or 10 mg/kg, although exhibits neuroprotection in transient global forebrain ischemia, the extracellular glutamate level during ischemia is not a major determinant of the neuroprotective activity of propofol.[17] Furthermore, the glutamate receptor antagonists have been shown to protect rodents and cats from cerebral injury in both focal [18,19] and transient global ischemia [20] models.

There are many more areas of interest for translational research in neurosciences like cognitive dysfunction, cerebral function monitoring, traumatic brain injury, pharmacogenomics etc., where translational research has immense scope, which may further help find answers to queries related to these fields. Postoperative cognitive dysfunction is another unexplored research area where a good quality translational research can be planned, which may further help find an answer to queries related to cognitive functions and may help understand the conditions like aging, dementia, and Alzheimer's disease.[21,22] In the neurointensive care as well as in operation theaters, the need and the importance to preserve neuronal viability demands research into good technologies that have better sensitivity and specificity. In acute brain injury like traumatic brain injury, in comparison to primary insult, secondary insults play a significant role in increasing the morbidity and mortality of the patients. As of now, there is no definitive therapy that restores the neuronal integrity after traumatic brain injury and that calls for translational research with an aim to prevent secondary injury. With regard to translational research in *cerebral function monitoring*, its need of the hour since the existing monitors are cumbersome, expensive, invasive with limited sensitivity and specificity. The importance to preserve neuronal viability during surgery and also in the neuro-intensive care units demands research into newer, innovative non-invasive, inexpensive technologies that have better sensitivity and specificity. If we talk about *pharmacogenomics*, the study of the effect of drugs based on the individual genetic profile of the patient, in recent years there has been evidence that anesthetic drugs, their action, and their side effects are dependent on genetic polymorphism. It has been observed that genetic factors do contribute to a majority of severe adverse drug reactions.[23] Days are not far when anesthetics will be chosen on the basis of the genetic profile of the patient in order to avoid any unwanted side effects.[24] About mechanical ventilation in patients with neurological diseases like Guillain–Barre syndrome and myasthenia gravis, research done during 1950s European polio epidemics may be considered a pioneering effort at translational science that paved the way for mechanical ventilation to save lives of patients with reversible neuromuscular diseases.[25,26] The successful ventilation strategies in these patients in present times in neurological practice are only because of the translational work of those early years.

In conclusion, we can say that translational research has improved the efficiency of health care providers in past. In the future, there will be much more progress in this field which will help medical research to excel beyond its scope and achieve a new height.

References

1. Butler D. Translational research: crossing the valley of death. *Nature*. 2008;453: 840–842.
2. Collins FS. Reengineering translational science: the time is right. *Sci Transl Med*. 2011;3. 90cm17.
3. Davidson A. Translational research: what does it mean? *Anesthesiology*. 2011;115: 909–911.

4. Michenfelder JD, Gronert GA, Rehder K. Neuroanesthesia. *Anesthesiology*. 1969;30: 65–100.
5. Khan FA. Translational research and anesthesia. *J Anaesthesiol Clin Pharmacol*. 2014;30:151–152.
6. Loepke AW, Istaphanous GK, McAuliffe III JJ, et al. The effects of neonatal isoflurane exposure in mice on brain cell viability, adult behavior, learning, and memory. *Anesth Analg*. 2009;108:90–104.
7. Satomoto M, Satoh Y, Terui K, et al. Neonatal exposure to sevoflurane induces abnormal social behaviors and deficits in fear conditioning in mice. *Anesthesiology*. 2009;110:628–637.
8. Fredriksson A, Ponten E, Gordh T, Eriksson P. Neonatal exposure to a combination of N-methyl-D-aspartate and γ-aminobutyric acid type A receptor anesthetic agents potentiates apoptotic neurodegeneration and persistent behavioural deficits. *Anesthesiology*. 2007;107:427–436.
9. Viberg H, Ponten E, Eriksson P, Gordh T, Fredriksson A. Neonatal ketamine exposure results in changes in biochemical substrates of neuronal growth and synaptogenesis, and alters adult behavior irreversibly. *Toxicology*. 2008;249:153–159.
10. Jevtovic-Todorovic V, Hartman RE, Izumi Y, et al. Early exposure to common anesthetic agents causes widespread neurodegeneration in the developing rat brain and persistent learning deficits. *J Neurosci*. 2003;23:876–882.
11. Ben-Ari Y. Excitatory actions of GABA during development: the nature of the nurture. *Nat Rev Neurosci*. 2002;3:728–739.
12. Slikker Jr W, Zou X, Hotchkiss CE, et al. Ketamine induced neuronal cell death in the perinatal rhesus monkey. *Toxicol Sci*. 2007;98:145–158.
13. Ikonomidou C, Bosch F, Miksa M, et al. Blockade of NMDA receptors and apoptotic neurodegeneration in the developing brain. *Science*. 1999;283:70–74.
14. Wells BA, Keats AS, Cooley DA. Increased tolerance to cerebral ischemia produced by general anesthesia during temporary carotid occlusion. *Surgery*. 1963;54:216–223.
15. Michenfelder JD, ed. *Anesthesia and the Brain*. New York: Churchill Livingstone; 1988:35–48.
16. Schwer CI, Lehane C, Guelzow T, et al. Thiopental inhibits global protein synthesis by repression of eukaryotic elongation factor 2 and protects from hypoxic neuronal cell death. *PLoS One*. 2013;8:e77258.
17. Yano T, Nakayama R, Ushijima K. Intracerebroventricularpropofol is neuroprotective against transient global ischemia in rats: extracellular glutamate level is not a major determinant. *Brain Res*. 2000;883:69–76.
18. Park CK, Nehls DG, Graham DI, Teasdale GM, McCulloch J. Focal cerebral ischaemia in the cat: treatment with the glutamate antagonist MK-801 after induction of ischaemia. *J Cereb Blood Flow Metab*. 1988;8:757–762.
19. Sarraf YS, Sheng H, Miura Y, et al. Relative neuroprotective effects of dizocilpine and isoflurane during focal cerebral ischemia in the rat. *Anesth Analg*. 1998;87:72–78.
20. Gill R, Foster AC, Woodruff GN. MK-801 is neuroprotective in gerbils when administered during the post-ischaemic period. *Neuroscience*. 1988;25:847–855.
21. Fodale V, Ritchie K, Rasmussen L, Mandal PK. Anesthetics and Alzheimer's disease: Background and research. *Preface J Alzheimers Dis*. 2010;22(1-3):57.
22. Fodale V, Santamaria LB, Schiffilliti D, Mondal PK. Anaesthetics and postoperative cognitive dysfunction: a pathological mechanism mimicking Alzheimer's disease. *Anaesthesia*. 2010;65:388–395.

23. Chidambaran V, Ngamprasertwong P, Vinks AA, Sadhasivam S. Pharmacogenetics and anesthetic drugs. *Curr Clin Pharmacol*. 2012;7:78–101.
24. Mikstacki A, Skrzypczak ZM, Tamowicz B, Zakerska BO, Szalata M, Slomski R. The impact of genetic factors on response to anaesthetics. *Adv Med Sci*. 2013;58:9–14.
25. Sund KH, Lunding M. Two early Danish respirators designed for prolonged artificial ventilation. *Acta Anaesthesiol Scand Suppl*. 1978;67:96–105.
26. Haglund G. Respiratory treatment in polio epidemics 1953-1961. *Acta Anaesthesiol Scand Suppl*. 1962;12:25–29.

Basic principles of translational research

2

Yusuke Naito and Masahiko Kawaguchi
Department of Anesthesiology, Nara Medical University, Kashihara, Japan

Introduction to translational research

The COVID-19 outbreak that was reported at the end of 2019 rapidly caused a pandemic, resulting in many hospitalizations and halting economic activity. Although governments and health organizations have proposed effective measures as of 2020, including limiting social activities (e.g., maintaining a social distance and locking down cities), prevention methods (e.g., face masks and face shields), and therapeutic agents (e.g., remdesivir and dexamethasone), the pandemic has yet to be resolved. Under these circumstances, the development of an effective vaccine has become of primordial importance, but this development can take many years, up to a decade, before it can be released to the market. However, due to high social demands and large research investments, some countries approved compassionate distribution by the end of 2020 and several countries expected to introduce the vaccine by 2021. Research that begins on the basis of such high social demands, and that will contribute to actual clinical practice and society, is called translational research (TR). Morris et al. noted that it takes an average of 17 years for a basic research project to transition through clinical research and be fully translated into day-to-day clinical practice, and that many of these projects are buried along the way without ever being adopted into daily practice.[1] Therefore, the ultimate mission of TR is to clarify and remove obstacles, and shorten this lengthy study/application period. The purpose of this chapter is to outline the existing TR and overview its history, current concepts, ethical issues, and education to determine the gaps within the studies and effectively promote the significance of such research.

What is translational research?

According to the current definition, TR is "a series of studies starting with questions arising from clinical practice, followed by basic research, and ending with research that provides benefits to society." This broad definition encompasses a wide array of processes and research, including basic research, animal testing, propagating

Perioperative Neuroscience. DOI: https://doi.org/10.1016/B978-0-323-91003-3.00013-1

research results, research focused on human application, comparisons with existing treatments, clinical guideline development, collaborations with manufacturers, epidemiological studies, promoting patients' behavioral changes, reflecting research results in policy, and stimulating research in peripheral areas. In this section, we describe TR using Gruentig's research in coronary catheterization, one of the most famous examples of TR.

Currently, the standards of care for coronary artery stenosis are either percutaneous coronary intervention (PCI) or coronary artery bypass graft (CABG), the latter of which emerged first. In 1946, Veinberg reported on the endocardial implantation of the internal thoracic artery in 88 patients[2] and in 1967, Favaloro et al. successfully performed a direct anastomosis of the coronary arteries using the interposition method with the great saphenous vein graft, a method that is still used today.[3] However, at that time, only the on pump CABG was performed, as knowledge on postoperative multidisciplinary management was lacking, and this frequently ended with complications and deaths, due to excessive surgical stress. Gruentzig, a former radiologist in Germany, tried to apply the concept of catheterization for atherosclerosis obliterans, proposed by Zeitler et al., to the treatment of coronary artery stenosis. After completing his clinical work, Gruentzig worked with his assistant and family in his kitchen to develop a high-pressure, low-volume balloon that could be used to open coronary arteries. Subsequent animal studies showed that coronary artery patency was possible and this originated the current percutaneous old balloon angioplasty, representing the first step of TR, from bedside to bench. He presented the results at the 50th American Heart Association Scientific Session and actively gave live demonstrations, teaching the technique to physicians all over the world, which quickly spread the method (results transmission).

In 1982, a steerable guidewire was developed in collaboration with the manufacturer to improve the difficulties clinicians experienced in ensuring that a guidewire reaches the targeted lesion, resulting in the catheter's distal advancement (collaboration with vendors). In 1987, after sufficient cases were collected to validate its superiority over CABG, the National Institute of Health (NIH) led a comparative study of percutaneous transluminal coronary angioplasty and CABG (comparison with existing treatments).[4] The bare metal stent was introduced in 1987 to overcome the problem of restenosis, and low-profile, long, and high-pressure balloons, as well as perfusion and monorail devices, were developed one after another in response to clinical demands. Then, in 2004, the Drug Eluting Stent was introduced and it improved the long-term outcomes. On the other hand, in response to the advent and improvement of PCI, the off pump CABG and other devices, such as the automatic anastomosis, became popular from the mid-1990s onward, as they reduced the risk of cerebral infarction, due to ascending aorta clamping. In addition, new techniques, such as the minimally invasive direct CABG, emerged to avoid mediastinitis (research stimulation in peripheral areas). The development of these treatments for coronary artery stenosis is a classic example of TR, in which a single radiologist brings the bedside to the bench, conducts basic research, collaborates with manufacturers, widely disseminates knowledge, conducts clinical research, develops guidelines, encourages improvements in peripheral areas, and returns some of these improvements into basic research.[5] In other words, TR begins with a clinical problem and produces basic research, whose final results then penetrate into and give back to society.

History of translational research

The term "TR" first appeared in biomedical journals in 1993. At that time, it was mostly used in cancer research,[6,7] because the Specialized Programs of Research Excellence (SPOREs), led by the National Cancer Institute (NCI) and initiated in 1992, addressed the United States society's lack of clinical applications of basic research results in tumor prevention and treatment, and encouraged the designing and performing of clinically relevant basic research.[8] The NCI was the first to use "from bench to bedside" in the SPOREs and this phrase is still frequently used to describe TR.[9] Furthermore, the term "TR" became famous when Dr. Pinedo of Free University, who received the Steiner Award in 1995, gave an award lecture entitled "Drug resistance, illustration of the complexity of TR."[9] Highlighting the divergent goals of the following two processes, Pinedo pointed out that in clinical practice, highly invasive treatments, such as anticancer drugs and surgery, are used once cancer has progressed, while in basic research, resources are devoted to the dissection of oncogenes and the mechanisms of cancer development. He also noted that the development of drugs, such as tamoxifen and retinoids, that are effective against early-stage cancers, and research on peptides, such as the gastrin-releasing peptide, which is a biomarker for early cancer detection, are both important for improving public health, rather than treating cancer after it has progressed. To achieve this goal, TR should combine both clinical and basic science perspectives.[10]

In 2006, the NIH established the Clinical and Translational Science Funding (CTSA) program, providing $4–$9 million per site and leading to the creation of many research facilities in the United States bearing the TR name. With the addition of the NIH, TR became widespread in areas other than oncology, and TR spread mainly to AIDS,[11] cardiology,[12] and gastrointestinal endocrinology research.[13] The UC Davis Clinical and Translational Center, and the University of Pittsburgh's Clinical and Translational Science Institute are representative examples of TR centers in the United States that the CTSA operates. Outside the United States, the concept of TR spread to the SIBS-Novo Nordisk Translational Research Center in Shanghai, the University of Cambridge, the University of Newcastle, the Imperial College of the University of Newcastle, the Imperial College of London, and the consortium of Scottish Institutions in the early the 2000s. In the latter half of the 2000s, not only academic and educational institutions but also pharmaceutical companies and medical device manufacturers began to focus on TR, including both pre- and post-commercialization examinations into the concept of TR. For example, the Wellcome Trust is a joint fund established by major pharmaceutical companies, such as GlaxoSmithKline, Pfizer, Roche, AstraZeneca, Sanofi, Aventis, and Sirtris Pharmaceutical. Also, the University of Cambridge established the Translational Medicine and Therapeutics program to promote TR.

As the focus on translational medicine increased, its originally vague definition, a bridge between basic and clinical medicine, gradually expanded to various interpretations and definitions. For instance, with the participation of pharmaceutical companies, collaborations in industry and clinical trials are now included in the definition of translational medicine. In addition, with the funding of national organizations, such as the NIH, the development of guidelines and public behavioral studies were classified as TR in a broad sense. Thus, over the past 30 years, tTR which began spreading in the late 1990s, has evolved from a narrow definition of basic research leading to the discovery of new drugs and biomarkers to a broader definition: basic research that is based on clinical questions and can be translated into clinical practice. In 2000, the first International TR and preclinical Strategies in Clinical Radio-oncology Conference was held in Switzerland, and since then, there have been continuous TR conferences in many fields. Currently, there are 94 journals that accept submissions and are listed in PUBMED with the word "translational" in the journal name, many of which coincide with the events of the 2000s (Fig. 2.1).

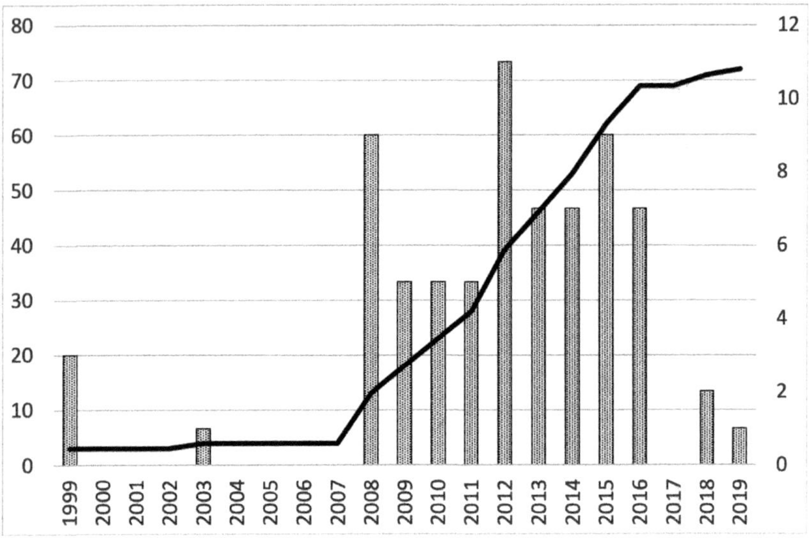

Figure 2.1 Trends in the number of journals with "TR" in their name. *Note*: The bar graph shows the number of new journals published in the year and the line graph reveals the total number of journals.

Current concepts and domains

In the 1990s, the Institute of Medicine Clinical research round table (now the National Academy of Medicine) classified TR into two phases: T1 (basic research until the first application on humans) and T2 (all subsequent processes).[14] Table 2.1 describes TR's classical classifications and the phases of clinical trials for newly approved drugs (Table 2.1). However, as TR has become more complex and encompasses a variety of concepts, a more detailed classification is now available.[15,16] In this section, we describe Khoury et al.'s definition in detail, although there are many other definitions and even similar research is categorized into different phases, resulting in no unified definition and a continued confusion. Table 2.2 summarizes the classification of typical TR. The word "phase" is also used in clinical trials for new drugs, classified into Phases 1–4. To avoid confusion, this document uses the letter T to distinguish between phases in TR, in accordance with past literature.

Table 2.1 Classification of clinical trials for new drug development and classical TR.

Traditional TR	New drug development	Phase goals	Drug doses	Number of recruited individuals
T1	Phase 0	Exploratory research into whether and how new drugs and new methods actually work	Small amount (below the effective threshold)	A few healthy people
	Phase 1	Detection of serious side effects. Conducting pharmacokinetic studies	Up to Normal dose (Step up manner)	20–30 patients
	Phase 2	Evaluating drug efficacy	Dose determined in Phase 1	25–100 patients with specific cancers
T2	Phase 3	Comparisons with current treatment	Dose for Phase 2	Hundreds of people in the scale
	Phase 4	Detection of rare complications. Conducting cost-effectiveness studies.	Post-marketing (packet insert)	Thousands

Table 2.2 Primary definition labeling of TR.

	Sung[17]	Westfall[18]	Woolf[10]	Dougherty[19]	Khoury[20]
Biomarker discovery	T1		T1	T1	T1
Target development					
Preclinical development					
Phase 1	-	T1	-		
Phase 2					
Phase 3				-	T2
Guideline formation	T2	T2	T2		
Phase 4	-	-	-		T3
Comparative effective research				T2	-
Health service research		T2			
Implementation research	T2		T2	T3	T3
Dissemination research	-	T3	-		
Local outcome research		-		T2	T4
Population level outcome research					
Cost effectiveness research				-	
Population research					

Note: "-" indicates not mentioned in the publication.

Phase T1 in TR, as defined by Khoury et al., includes the process from basic research all the way through to the early phases of clinical trials on humans. This includes research in drug development, proteomics, genomics, genetics, metabolomics, and animal modeling. Other examples are the development of diagnostic devices and modalities, as well as Phase 1 and 2 clinical trials. Phase T1 in TR represents the "bench to bedside" process and is roughly equivalent to the T1 in the classical classification. Phase T2 in TR includes early-stage studies on the efficacy and human safety, and the development of clinical guidelines. Phases 3 and 4, representing the clinical trials of new drugs as well as comparative studies with existing therapies and interventions, are also examples of Phase T2 research. Lastly, comparative effectiveness research, which includes cost analyses, has been proposed and conducted extensively in recent years. The related studies are also included within the T2 phase.[21] While T1 is called "bench to bedside," phase T2 in TR is described as "bedside to practice."

Phase T3 in TR includes Phase 4 clinical study and dissemination and implementation research (D&I research) and refers to research that focuses on the dissemination of Phase T2 results into clinical practice. D&I research, a commonly used term in the United States, aims to develop and validate methods to effectively and efficiently incorporate EBI into daily health and welfare activities, and build a knowledge system. In Canada, the term "knowledge translation" replaces D&I, but refers to the same process.[22] Moreover, phase T4 in TR includes studies that focus on outcomes and effectiveness at the community or entire population level, and public health and epidemiology play a central role. Specifically, this phase includes long-term cohort studies at the municipal level, the results of which are often used to determine health policy, identify problems, and return to basic research, forming a closed TR loop.

Although TR is subdivided into four phases, it is impossible to exhaustively categorize all the studies and there are barriers, called gaps, between the phases. To overcome these gaps and propagate data to the next study, data transferring is important.[23] Throughout the study process, the number of researchers to whom the data is transferred increases as the phase progresses. For example, at the very beginning of T1, basic data is only shared among basic researchers, but as the phase progresses, animal experimenters and biochemists are gradually involved, and in the later phases, practitioners, such as policy makers and pharmaceutical companies, are included. The research results are typically shared within communities of researchers in the form of publications and presentations at scientific meetings. This approach carries the risk that the information is only being shared among researchers within the same profession, and not among other professionals, such as clinicians and policy makers. Therefore, it is important that data transferring is explained and publicized in a way that makes it possible for people who do not have research expertise to interpret the results. One of the best and most concrete examples of this is the Cochrane systematic review that requires a plain-language summary of the results to ensure that nonspecialists can interpret them. Although this approach facilitates data transfers, there is a potential risk that exceptional events may be overly simplified, as the results are generalized when presented as a summary.[15]

Other than data transfer failures, translation errors can also occur. In particular, many studies have been terminated without proceeding to T2, due to the gap between the T1 and T2 phases, called the "valley of death."[24] According to Pienta et al., there are five risks in this valley of death: funding, scientific, intellectual property, market, and regulatory risks. The scientific risk depends on the strength of the scientific support at T1, while intellectual property risk relates to the discovery's novelty, existing patents for prior experiments, and competing research. As for regulatory risks, they tend to vary among countries. For instance, in the United States, a New Drug Application must be submitted to the FDA at the end of phase 3 to obtain approval for general use. To overcome this valley of death, some universities, such as the University of Michigan, not only provide financial support for TR, but also review the research design with expert committee members, verify whether the research infringes others' patents (patent risks), consult with economic experts from the business community to predict the commercialization potential, and advise on new drug applications with the help of legal experts.[23]

In addition, the T2 and T3 gap refers to situations in which effective therapies are established in T2, but not introduced into clinical practice or for the general public. For example, using beta-blockers after an acute myocardial infarction is known to be effective and more beneficial than developing new drugs, with high-cost benefits. However, in practice, only 60% of eligible patients receive this treatment.[25] The T4 study included behavioral aspects of the treatments that were validated in T3, but the T3 results did not necessarily lead to the T4 study. As another example, studies on the benefits of statins have shown that obese patients and smokers are at risk for hypercholesterolemia and cardiovascular disease, but not all patients follow the recommended behavioral changes, such as exercising and quitting smoking. As such, developing new drugs for high cholesterol (T1 studies) remains highly competitive. Overall, by classifying TR into phases and discussing the gaps, it is possible to determine the shortest and most effective method depending on the research focus.

Ethics and regulations

Bioethics is an important indicator for weighing the rights of an individual against public welfare. Although freedom of scientific inquiry is essential for medical progress, abuses of rights in medical research, such as the injection of live cancer cells into patients at the Jewish Hospital for Chronic Diseases in Brooklyn and the Tuskegee syphilis experiment, indicate that the right to scientific inquiry is not absolute. Table 2.3 lists the international guidelines that must be followed in all phases of TR. Although it is essential to conduct research in accordance with these guidelines, the interpretation and application of the guidelines vary according to each country's cultural background and ideologies, which may affect the research process. Indeed, obtaining informed consent is the most important procedure in clinical research recruitment and represents a patient's right. However, organ transplants and related research, for instance, require living consent in Germany, while Austria has an opt-out system. Although Austria and Germany are culturally similar, the consent rate differs greatly from 12% to 100%, due to differences in the methods for obtaining consent.[26,27] Thus, many reviews have been published in recent years on adapting the principle theory of bioethics.[14,28] There have been cases in which a study that was successfully becoming a form of TR was halted because bioethical issues could not be overcome and conversely, there have been other cases in which bioethics led to the development of other types of research. In the following subsections, we describe examples of ethical issues that exist at each stage of TR and discuss how these issues can be addressed.

Table 2.3 International ethical guidelines and conventions.

Nuremberg Code (1947)	During the Second World War, many unethical human experiments were conducted in concentration camps by researchers who were complicit with Nazi Germany. In response, the Nuremberg Code of Conduct was compiled as an ethical guideline for human experimentation in medical research. This Code, which clearly indicates that the individual's voluntary consent is essential for experimentation, is still repeatedly referred to today as the prototype for ethical guidelines.
Declaration of Geneva (1948)	Describes the ethical principles that modern doctors should follow to fulfill their duties. In 1948, the World Medical Association adopted the guideline, calling it the modern version of the Hippocratic Oath. The guideline was periodically revised in 1968, 1983, 1994, and 2006, which lead to the current guideline.
Declaration of Helsinki (1964)	In 1964, the World Medical Association adopted the Declaration as an ethical guideline for clinical research on humans, based on the Nuremberg Code. In 1975, it was revised to introduce the term "informed consent," an important concept in bioethics.
Summary Statement of the Asilomar Conference on Recombinant DNA Molecules (1975)	In the later half of the 20th century, advances in molecular biology made it possible to artificially recombine certain genes. In 1975, scientists, who felt threatened by this discovery, held a conference in Asilomar, USA, and established guidelines to avoid biohazards. This is regarded as an epochal event because scientists recognized and voluntarily addressed the dangers of experimentation.
Belmont Report (1979)	This report represents a more thorough examination of the ethical principles raised in the Nuremberg Code and a consistent systematization of various concepts. It offered the three universal principles of "respect of persons,"

(*Continued*)

Table 2.3 (Continued)

	"beneficence," and "justice." It also clearly defined research activities and stipulated those subject to ethical review, turning it into an extremely strict guideline.
Convention on Biological Diversity (1992)	The United Nations adopted this convention as an international treaty for the conservation of the diversity of life on Earth. It is an epochal treaty in that it serves as a guideline for the conservation of not only specific organisms or regions but also of biological diversity in general. Another important feature of the Convention is that it emphasizes the "sustainable use" of such biological resources.
Universal Declaration on the Human Genome and Human Rights (1997)	In response to the Human Genome Project that various countries launched in the early 1990s, this declaration established basic guidelines aiming to ensure that the human genome research results are not used to violate human dignity and rights or to discriminate against people in society while affirming them as a "legacy of humanity."
Proposed International Guidelines on Ethical Issues in Medical Genetics and Genetic Services (1998)	This is an ethical guideline for genetic medicine, proposed by the WHO. It is based on the principles of respect for the individual's right to self-determination, individual welfare, and well-being, while also referring to issues that are specific to genetic medicine, such as information management, and considerations for families and ethnic groups.
International Ethical Guidelines for Health-related Research Involving Humans (2002)	Ethical guidelines proposed by the International Council of Organizations for Medical Sciences. The most important feature is the provision of guidelines for the proper conduct of international collaborative research that also considers the problems of infectious diseases occurring in developing countries. The guidelines are of great importance, as they indicate the type of medical research that is required in this era of globalization.

(*Continued*)

Table 2.3 (Continued)

International Declaration on Human Genetic Data (2003)	The declaration highlights that genetic information can be used to predict individual diseases, can affect family members and other groups, and contains details of unknown significance at the time of sample collection.
Universal Declaration on Bioethics and Human Rights (2005)	Adopted by UNESCO. This declaration consists of 15 items, one theme being how to organize the diversity of cultures and the universality of principles.

Ethical issues in T1: embryonic stem cells and induced pluripotent stem cells

Advances in organ transplantation have made it possible to harvest many organs, such as the heart, liver, and kidneys, from the living or deceased and replace them with failing organs. While many patients could potentially benefit from this treatment, there is a limited number of available organs that can be removed. As embryonic stem cells have the ability to transform into all cells within an organism, they were expected to be logistically capable of producing organs or even entire human bodies.[29] However, because their production requires the destruction of a human embryo, ethical issues arose. Specifically, a series of studies raised two ethical issues: whether embryos can be considered living and whether cloned individuals should be treated ethically.

However, induced pluripotent stem (iPS) cells partially solved this problem. Yamanaka et al. succeeded in establishing human iPS cells by introducing cells into fibroblasts that were obtained from human skin (cells without differentiation potential).[30] When injected subcutaneously into immunodeficient mice, the iPS cells grew into teratomas, consisting of endoderm (intestinal-like epithelium), mesoderm (skeletal muscle and cartilage), and ectoderm (neural tissue and epidermis), within nine weeks, indicating that they could potentially produce organ parts.[31] Thus, ethical guidelines not only serve as barriers but also have the potential to steer research in a more ethical direction.

Ethical issues in T2: artificial red blood cell products

One recent example is the Northfield Laboratories-sponsored PolyHeme multi-institutional study[32] in which the clinical trial protocol for the drug was designed to ensure the randomization of eligible patients with hemorrhagic shock, receiving either saline or PolyHeme, a type of artificial red blood cell preparation. The participants assigned to the intervention group received up to six units of PolyHeme, instead of red blood cells, for up to 12 hours. Once the allotted time had elapsed or six units of PolyHeme had been administered, the participants in the experimental group received a regular blood transfusion, as needed. This protocol was criticized for not obtaining enough informed consent and its lack of treatment choices with established safety, due to the nonexistence of an adequate transfusion supply system for hemorrhagic shock in the United States. As a result, institutional review boards (IRBs) at some facilities withheld the approval of the PolyHeme study and it was ultimately suspended. PolyHeme was expected to be a safe and long-lasting alternative to blood transfusion, but it had to be suspended for ethical reasons and the Northfield Labs withdrew from the project in 2009. Since 2006, another artificial red blood cell product, Hemopure, has commercially been used in South Africa,[33] as the risk of infection from blood transfusion products is still high in the country and Hemopure's benefits outweigh the potential side effects. Thus, it is important to understand that ethical interpretations of similar situations may differ depending on environmental factors.

Ethical issues in T3–4: the thalidomide drug incident

A German pharmaceutical company launched thalidomide, as a nonbarbiturate, nonaddictive sedative, but later, when it was discovered that it prevented nausea, it was prescribed for pregnant women's morning sickness. It was recognized as a safe drug with very few side effects and was frequently used in 46 countries, but the United States FDA stopped marketing the drug because they could not confirm its safety for pregnant women.[34] The first reported side effect was peripheral neuropathy and although many patients complained, pharmaceutical companies and various countries' regulatory authorities ignored this issue, because they lacked the ethical standards to subsequently collect and publish this information. Later, more than 1,000 deformed babies were born to pregnant women who took thalidomide, making it the biggest drug-related incident in the world.[35,36] In 1964, four countries, including the United Kingdom and the United States, launched adverse drug reaction monitoring systems and in 1968, the WHO established an international adverse drug reaction monitoring system. As with other drugs at the time, the safety of thalidomide was evaluated using rodents, such as mice, but the effects of species differences were significant and teratogenicity was solely observed in humans. This led to the need for new drugs to be designed in a way that considers species and racial differences in safety evaluation studies.

As shown in the above three examples, ethical issues in TR are composed of many perspectives, and it is difficult for a single researcher to understand and comply with all regulations and laws. One solution is to use team science to clarify the ethical issues that need to be addressed by involving a multidisciplinary team from the early stages of research, and receiving advice and guidance from other researchers. Another approach is to utilize research ethics consultation (REC), which has gained traction in recent years. The main role of REC is to advise on research design, conduct workshops on ethical issues, and assist in monitoring research violations.[37] Although IRBs have similar functions, RECs can provide consultation at any research stage (e.g., planning, conducting, interpreting, and disseminating), and approvals and reviews at the start of research. Of the 46 institutions funded by CSTA, 33 (70%) have REC services and more are expected to follow.[38]

Education and careers

TR, as described in previous sections, refers to a series of studies that are based on a basic research idea and will ultimately contribute to public health and welfare. Physicians, basic researchers, bioengineers, patent experts, statisticians, epidemiologists, policy makers, product companies, and other professionals are all involved in the research process. As untrained individuals cannot comprehensively coordinate all the processes of TR, it is important to involve experts who can coordinate these processes. Also, due to the variety of professions involved, it is necessary to at the least have a platform in which all researchers can communicate with other professionals using a common language.

On that note, Ph.D. scientists and MD (physician scientists) play a central role in TR. In the United States, the number of MDs has more than doubled to 900,000 in the past 30 years, but the number of physician scientists has remained steady at about 15,000.[39] The most important issues for physicians aiming to conduct TR are time, cooperation, and resource availability (e.g., laboratories and equipment at the physicians' facilities), as well as appropriate expert availability. Physicians need to spend time outside of their clinical practice to conduct research, but funding tends to be scarce because these individuals are a major source of revenue for hospitals. In addition, many institutions do not have sufficient specialists or departments to conduct TR, and it takes a great deal of time and effort to conduct research on one's own. In recent years, advances in medicine have led to an increase in clinical education learning that has, in turn, put pressure on the basic research curriculum. On the other hand, although scientists have sufficient time and knowledge to conduct TR, they are unaware of the ways in which their research will reflect in clinical practice, increasing the risk that they will pursue research that is unlikely to be translated into actual clinical practice. Thus, professionals who hold MDs and PhDs both require further education to understand how to conduct TR, but each has different deficiencies, suggesting the necessity of separate educational programs.

Currently, there are 90 available programs for training and forming physician scientists in the United States, with approximately 5,400 students in these programs. Typical programs include pathology, genetics, neuroscience, biomedical engineering, immunology, cancer biology, biochemistry, statistics, pharmacology, and other fields in basic science. Many of these programs are designed to supplement the basic medical knowledge that was unavailable when the individuals were obtaining their MDs. Although these programs are open to physicians at any point in their careers, the number of required training years before entering a PhD program has increased in recent years, partly due to financial problems. In the United States, the medical scientists training program of the

National Institute of General Medical Science funds about half of the programs, and although scholarships are usually provided, the funds are simply insufficient compared to clinicians' incomes. As another current problem, it is now more difficult to develop one's career as a physician scientist than in the past decades. In the 1970s, it took about three years to acquire a full-time job after obtaining a degree, while in 2003, this span increased to six years.[40] In the United States, it is common to accept and utilize overseas researchers with MDs to supplement the shortage of physician scientists, but drastic reforms are needed to fill these positions.[41]

On the other hand, the Med-Into-Grad program, initiated by Howard Hughs Medical Institute in 2005, is famous for educating basic scientists to obtain their PhDs. This program, funded with $10 million, was created to expose basic graduate students to medicine and to train future leaders in TR.[42] For example, the University of California, one of the initial 13 institutions involved in the program, established the Integrating Medicine into Basic Science course for graduate students in their second year or later. As graduate students have already decided on their degree projects, they can interact with researchers from diverse backgrounds, while keeping their own research area as a base. These graduate students come from diverse backgrounds, such as biomedical engineering, molecular biology, chemistry, and nutrition, but all must spend five weeks in a clinical setting at the start of the program, because this process will give them an understanding of the realities of the field where their research will ultimately be applied. For example, the cardiovascular course includes access to cardiovascular surgery units, cardiac catheterization laboratories, echocardiography, and heart failure outpatient clinics. The program as a whole also involves direct readings of the literature, patient management conferences, as well as courses on molecular mechanisms of disease, team dynamics, and team management.[43] Lastly, there are also several programs for nongraduate education.

With regards to other programs, the National Center for Advancing Translational Research provides webinars and promotes TR, and the European Advanced Translational Research Infrastructure in Medicine, which has 10 participating countries, is working on target validation, predictive models, biomarkers, clinical trials, IP, and European regulations. Thus, although expert training in TR has begun in many countries, the number of experts is rather small and the field's expansion is anticipated.

Conclusions

This chapter outlined the basic principles of TR that include a vast array of research methods and researchers, in which basic research is initiated based on clinical needs that are subsequently offered to society as resolutions to a problem. For the effective promotion of TR, it is important to understand the gaps that exist between each study (Table 2.4).

Table 2.4 Nuremberg code.

1. The voluntary consent of human participants is absolutely essential. This means that persons involved should have the legal capacity to give consent; be able to exercise free power of choice, without the intervention of any type of force, fraud, deceit, duress, over-reaching, or other ulterior forms of constraint or coercion; and have sufficient knowledge and comprehension of the elements involved in the subject matter, to enable him/her to make an informed and enlightened decision. The latter requires that before accepting the studied persons' affirmative decisions, the nature, duration, and purpose of the experiment; the method and means by which it is to be conducted; all reasonably expected inconveniences and hazards; and the potential health or personal effects that may result from participation should all be made known to the participant. The duty and responsibility for ascertaining the quality of the consent rests upon each individual who initiates, directs, or engages in the experiment. It is a personal duty and responsibility that may not be delegated to another with impunity.
2. The experiment should not be random and unnecessary in nature and should yield fruitful results for the good of society that are unprocurable by other methods or means of study.
3. The experiment should be designed based on the results of previous animal experimentation and a knowledge of the natural history of the disease or other problem under study. The anticipated results should justify the performance of the experiment.
4. The experiment should be conducted in a way that avoids all unnecessary physical and mental suffering and injury.
5. No experiment should be conducted when there is an a priori reason to believe that death or disabling injuries could occur, except, perhaps, in those experiments where experimental physicians also serve as participants.
6. The degree of risks should never exceed those determined by the humanitarian importance of the problem that the experiment aims to solve.
7. Proper preparations should be made and adequate facilities should be provided to protect the participants against even remote possibilities of injury, disability, or death.
8. Only scientifically qualified persons should conduct the experiment. Those who conduct or engage in the experiment must exercise the highest degree of skill and care throughout all stages of the experiment.
9. During the course of the experiment, the participants should be at liberty to quit if they reach a physical or mental state that makes them feel like continuing is impossible.
10. The scientist in charge must be prepared to terminate the experiment at any stage, if he/she has probable cause to believe, in the exercise of good faith, superior skill, and careful judgment, that a continuation is likely to result in participant injury, disability, or death.

Source: Adopted from https://history.nih.gov/display/history/Nuremberg + Code.

References

1. Morris ZS, Wooding S, Grant J. The answer is 17 years, what is the question: understanding time lags in translational research. *J R Soc Med*. 2011;104:510–520.
2. Vineberg AM. Development of an anastomosis between the coronary vessels and a transplanted internal mammary artery. *Can Med Assoc J*. 1946;55:117–119.

3. Cooley DA. In memoriam. Tribute to René Favaloro, pioneer of coronary bypass. *Tex Heart Inst J*. 2000;27:231–232.
4. King III SB, Lembo NJ, Weintraub WS, et al. A randomized trial comparing coronary angioplasty with coronary bypass surgery. Emory Angioplasty vs Surgery Trial (EAST). *N Engl J Med*. 1994;331:1044–1050.
5. King III SB. Angioplasty from bench to bedside to bench. *Circulation*. 1996;93:1621–1629.
6. van der Laan AL, Boenink M. Beyond bench and bedside: disentangling the concept of translational research. *Health Care Anal*. 2015;23:32–49.
7. Broder S, Cushing M. Trends in program project grant funding at the National Cancer Institute. *Cancer Res*. 1993;53:477–484.
8. Minna JD, Gazdar AF. Translational research comes of age. *Nat Med*. 1996;2:974–975.
9. Mulshine JL, Jett M, Cuttitta F, et al. Scientific basis for cancer prevention. *Intermed Cancer Markers*. 1993;72:978–983.
10. Woolf SH. The meaning of translational research and why it matters. *JAMA*. 2008;299:211–213.
11. Hunter E. Perspective: research highlights of the University of Alabama at Birmingham Center for AIDS Research. Paradigm shifts, translational research, and CFARs. *AIDS Res Hum Retroviruses*. 1996;12:1301–1303.
12. Block AJ. Translational research is hurting. *Chest*. 1999;115:311.
13. Rustgi AK. Translational research: what is it? *Gastroenterology*. 1999;116:1285.
14. Hostiuc S, Moldoveanu A, Dascălu MI, et al. Translational research-the need of a new bioethics approach. *J Transl Med*. 2016;14:16.
15. Waldman SA, Terzic A. Clinical and translational science: from bench-bedside to global village. *Clin Transl Sci*. 2010;3:254–257.
16. Trochim W, Kane C, Graham MJ, et al. Evaluating translational research: a process marker model. *Clin Transl Sci*. 2011;4:153–162.
17. Sung NS, Crowley Jr. WF, Genel M, et al. Central challenges facing the national clinical research enterprise. *JAMA*. 2003;289:1278–1287.
18. Westfall JM, Mold J, Fagnan L. Practice-based research–"Blue Highways" on the NIH roadmap. *JAMA*. 2007;297:403–406.
19. Dougherty D, Conway PH. The "3T" road map to transform United States health care: the "how" of high-quality care. *JAMA*. 2008;299:2319–2321.
20. Khoury MJ, Gwinn M, Yoon PW, et al. The continuum of translation research in genomic medicine: how can we accelerate the appropriate integration of human genome discoveries into health care and disease prevention? *Genet Med*. 2007;9:665–674.
21. Kirschner N, Pauker SG, Stubbs JW, et al. Information on cost-effectiveness: an essential product of a national comparative effectiveness program. *Ann Intern Med*. 2008;148:956–961.
22. Tabak RG, Khoong EC, Chambers DA, et al. Bridging research and practice: models for dissemination and implementation research. *Am J Prev Med*. 2012;43:337–350.
23. Pienta KJ. Successfully accelerating translational research at an academic medical center: the University of Michigan-Coulter translational research partnership program. *Clin Transl Sci*. 2010;3:316–318.
24. Butler D. Translational research: crossing the valley of death. *Nature*. 2008;453:840–842.
25. Lenfant C. Shattuck lecture–clinical research to clinical practice–lost in translation? *N Engl J Med*. 2003;349:868–874.
26. Ugur ZB. Does presumed consent save lives? Evidence from Europe. *Health Econ*. 2015;24:1560–1572.

27. Batlle D, Haque SK. Genetic causes and mechanisms of distal renal tubular acidosis. *Nephrol Dial Transpl.* 2012;27:3691–3704.
28. Mandal J, Ponnambath DK, Parija SC. Ethics of translational medical research. *Trop Parasitol.* 2017;7:62–64.
29. Biswas A, Hutchins R. Embryonic stem cells. *Stem Cell Dev.* 2007;16:213–222.
30. Watanabe A, Yamada Y, Yamanaka S. Epigenetic regulation in pluripotent stem cells: a key to breaking the epigenetic barrier. *Philos Trans R Soc Lond B Biol Sci.* 2013;368:20120292.
31. Takahashi K, Tanabe K, Ohnuki M, et al. Induction of pluripotent stem cells from adult human fibroblasts by defined factors. *Cell.* 2007;131:861–872.
32. Human haemoglobin - Northfield. *BioDrugs.* 2003;17:296–298.
33. Mer M, Hodgson E, Wallis L, et al. Hemoglobin glutamer-250 (bovine) in South Africa: consensus usage guidelines from clinician experts who have treated patients. *Transfusion.* 2016;56:2631–2636.
34. Vargesson N. Thalidomide-induced limb defects: resolving a 50-year-old puzzle. *Bioessays.* 2009;31:1327–1336.
35. Vargesson N. Thalidomide-induced teratogenesis: history and mechanisms. *Birth Defects Res C Embryo Today.* 2015;105:140–156.
36. Lenz W, Knapp K. Thalidomide embryopathy. *Dtsch Med Wochenschr.* 1962;87:1232–1242.
37. Sharp RR, Taylor HA, Brinich MA, et al. Research ethics consultation: ethical and professional practice challenges and recommendations. *Acad Med.* 2015;90:615–620.
38. McCormick JB, Sharp RR, Ottenberg AL, et al. The establishment of research ethics consultation services (RECS): an emerging research resource. *Clin Transl Sci.* 2013;6:40–44.
39. T L. *The Vanishing Physician-Scientist?* 1st ed. Cornell University Press; 2009.
40. Pickering CR, Bast Jr. RC, Keyomarsi K. How will we recruit, train, and retain physicians and scientists to conduct translational cancer research? *Cancer.* 2015;121:806–816.
41. Vidyasagar D. Integrating international medical graduates into the physician-scientist pool: solution to the problem of decreasing physician-scientists in the United States. *J Investig Med.* 2007;55:406–409.
42. Hutson S. HHMI's Med Into Grad Initiative expands. *Nat Med.* 2009;15:123.
43. Knowlton AA, Rainwater JA, Chiamvimonvat N, et al. Training the translational research teams of the future: UC Davis-HHMI Integrating Medicine into Basic Science program. *Clin Transl Sci.* 2013;6:339–346.

Anesthetics and translational research

3

Marco Cascella
Unit of Pain Medicine and Research in AnesthesiaIstituto Nazionale Tumori, IRCCS Fondazione Pascale, Napoli, Italy

Introduction

According to the classic triad of pillars "benchside," "bedside," and "community," translational medicine is a discipline in biomedical research that uses a multidisciplinary approach with a high level of collaboration, ranging from the research laboratory to the patient's bed and community.[1] This process should be understood in a bidirectional sense. Research should be applied to the patients from whom feedbacks on the applications of new treatments and possible improvements, are obtained.

The idea of this type of research approach was born in the last years of the 20th century when, despite an exponentially growing number of researches—and, therefore, also about the huge resources used—the results, in terms of outcomes improving, were struggling to arrive.[2] There was a need to apply basic research to the clinic and, vice versa, to direct preclinical research according to the needs dictated by potential clinical applications. At the same time, optimizing time and money resources was mandatory. Translational medicine, thus, works by combining different preclinical and clinical study areas, resources, and skills, to achieve improvements in the clinical practice in terms of prevention, diagnosis, and therapy development. The ultimate goal is to move in real-time from basic research fields such as in vitro and in vivo research to applied research (e.g., clinical research), to better apply the results in the clinical field.

This interdisciplinary research approach can be applied to all fields of medicine. The implementation of translational medicine programs in branches including but not limited to transplant medicine,[3] cardiology,[4] diabetes,[5] Alzheimer disease,[6] oncology,[7] pain therapy,[8] in fact, has led to significant clinical improvements. Ad hoc programs were also implemented during the COVID-19 emergency.[9]

In the field of anesthesia, translational medicine finds an important fertile ground. This discipline, indeed, practically intersects all biomedical branches. Furthermore, if the translation process must necessarily produce a health improvement, the outcomes of anesthesia are immediately evident, already in the intraoperative, but above all during and after emergence. Moreover, especially neuroscience can greatly benefit from the acquisitions produced by the study of anesthetics.

Perioperative Neuroscience. DOI: https://doi.org/10.1016/B978-0-323-91003-3.00008-8

Concerning the "bench", anesthesia can refer to basic research, even if it was aimed and developed for investigating in other areas, such as studies on brain functions, and the pathogenetic mechanisms of neurodegenerative diseases. On the contrary, scientific advances in the field of anesthesia and anesthetics can be used by other disciplines for the study of multiple phenomena.

This chapter is aimed at addressing some important aspects of translational research in anesthesia. For narrative reasons, the chapter has been divided into several paragraphs each addressing a specific topic. Although the list of topics related to translational research and anesthetics is particularly long, the reader's attention was focused on four main topics. These selected topics include:

- The mechanisms of anesthesia induction and emergence
- The anesthesia awareness phenomenon
- The relationship between general anesthetics exposure and neurodegenerative processes
- The impact of anesthetics on carcinogenic processes.

For each topic, an overview of preclinical and clinical findings and their applicability in clinical practice are discussed. Finally, research perspectives and an overview of the gaps to be overcome to enhance translational research in anesthesia are offered.

In this chapter, topics of fundamental importance such as the correlations between anesthesia/anesthetics and postoperative delirium, the phenomenon of emergence of delirium in children, or the potential neurotoxicity of anesthetics in the developing brain, the role of various antioxidants (vitamins, micronutrients, coenzymes) in conditions of brain damage, are not addressed. Nevertheless, although this lack represents an important limitation of the chapter, addressing few topics but more extensively discussed was considered a priority.

Induction and emergence from anesthesia processes

Overview on mechanisms

On the molecular level, the effect of general anesthetics is the result of their action on different types of receptors. Although the protagonists are gamma-aminobutyric acid type A ($GABA_A$) receptors, N-methyl D-aspartate (NMDA) receptors, and $\alpha 2$ receptors, other actors including but not limited to opioid receptors, ion channel anesthetic targets such as sodium and calcium channels, members of the neuronal hyperpolarization-activated cyclic nucleotide-gated family channels, and two-pore domain potassium channels, enter the scene. Receptors are not the only targets of anesthetics; especially inhaled anesthetics can bind to nonreceptorial cell structures including presynaptic release machinery, cytoskeleton elements, and mitochondrial complex I proteins. A particularly interesting, and useful data for translational research, is that anesthetics have targets that can be identified not only in humans but also in other organisms ranging from unicellular ciliates to plants, to primates. Functionally, the anesthetic binding to the receptor and/or to the cellular structures can produce perturbations in cortical and subcortical signals, inducing alterations in the connectivity across brain regions.

Data from preclinical research suggested that different general anesthetics interfere with different molecular patterns, and these processes can produce distinct functional alterations within the brain connectivity. In turn, the perturbations are manifested as different electrophysiological correlates. For example, volatile agents can impact the frontal-posterior connectivity by interfering with the gamma (20–60 Hz) oscillations which have a pivotal role in arousal and maintenance of consciousness[10] while propofol may induce a fast anteriorization of the alpha rhythms (8–12 Hz) also strengthening the propagations of slow-delta oscillations within the cortex. The final effect is a functional disruption of the connectivity between cortical areas. Furthermore, dexmedetomidine interferes with the thalamo-cortical functional connectivity—mostly expressed as spindle waves (12–16 Hz)—in the frontal area.[11]

Because the description of the precise neurobiology of the mechanisms of general anesthesia (GA) falls outside the scope of this chapter, we can briefly assume that, during induction, alterations in connectivity within distinct brain regions can produce different depths of anesthesia. Thus, changes in thalamic-cortical connectivity induce and accelerate the phenomenon of the loss of consciousness (LoC); similarly, perturbations in the cortico-cortical functional connectivity and a deeper effect on cortico-subcortical functioning are responsible for the completion of the induction mechanism and the maintaining of the surgical anesthesia status.

The ending stage of GA is featured by the progression from the patient's unconscious status to complete wakefulness and restoration of consciousness. This stage, called anesthesia emergence (AE), is characterized by neurobiology that largely differs from that of the induction.[12] During emergence, indeed, mechanisms responsible for LoC and anesthesia maintenance are gradually reversed. Nevertheless, these "passive" events are combined to ad hoc awakening mechanisms. These active processes provide the engagement of ascending arousal pathways mostly triggered by the thalamus[13] but also involving the substantia nigra, the ventral and laterodorsal tegmental areas of the midbrain, the dorsal raphe, and the locus ceruleus (LC) norepinephrine (LCNE) as well as the basal forebrain (BF), and lateral hypothalamus. Thus, it has been postulated the existence of a distinct mesencephalic arousal pathway. In particular, the LCNE is the main structure of the so-called LCNE arousal system which includes the posterior cingulate cortex, thalamus, and basal ganglia. The hypothalamus is also implicated in these wake-promoting processes as hypothalamic orexinergic neurons are involved in both the sleep-to-wake transition and maintenance of wakefulness.[14] This orexinergic system is functionally connected with other structures such as basal ganglia that regulate the awakening processes during the AE. Interestingly, serotonergic neurons in the dorsal raphe nucleus receive projections from orexinergic neurons, concurring in sleep–wakefulness modulation.[15] Another arousal pathway is the brainstem ascending reticular arousal system (ventral and dorsal pathways) which originates from the pontine/midbrain regions and develops cholinergic cortical projections through interactions with the thalamus, hypothalamus, and the BF region.

In summary, it can be highlighted that

- General anesthetics act on different types of receptors and nonreceptorial structures.
- Targets of anesthetics can be identified not only in humans but also in other organisms.
- During induction, alterations in connectivity within distinct brain regions can produce different depths of anesthesia.
- Changes in thalamic-cortical connectivity provoke and accelerate the LoC while cortico-cortical functional connectivity alterations and a further cortico-subcortical impairment induce the completion of the induction mechanism and the maintaining of the surgical anesthesia status.
- AE is not just a passive process due to a cessation of the action of general anesthetics.
- The arousal mechanisms at the end of GA are produced by structures deeper in the brain, rather than being activated within the neocortex.
- Rather than a single awakening system, it is more correct to refer to an arousal network.

Applicability in clinical practice

Since induction and awakening from anesthesia are asymmetric processes and, biologically, the neural circuits that mediate induction, at least in part, differ from those underlying the emergence, translational research has a great opportunity to investigate areas of extraordinary importance. This field, indeed, intersects different disciplines of neuroscience and can lead to incredible results in terms of patient outcomes. A better understanding of these processes has allowed researchers to start new studies focused on the actively induced transition from unconsciousness to awakening. Although the data are not conclusive, the prospects seem to be interesting.

Methylphenidate is a dopamine uptake inhibitor tested for this aim. Starting from the premises obtained from preclinical studies,[16] a randomized clinical trial (RCT) was conducted at Ohio State University (NCT02327195). In this investigation, methylphenidate (20 mg orally, 2 hours before surgery) was compared to placebo in adults undergoing isoflurane anesthesia. Although the recruitment status of this investigation is indicated as completed ($n = 54$), to date the results have not been published. Another RCT was conducted at Massachusetts General Hospital (NCT02051452). Patients were randomized to receive methylphenidate ($n = 16$) or placebo ($n = 16$). The study is listed as finished and we are waiting for published data. Another clinical study was scheduled to evaluate the drug in healthy volunteers, but it was closed prematurely (NCT02429076).

Another medication is modafinil. It is a long-acting wakefulness-promoting agent used for managing somnolence due to obstructive sleep apnea. Although its precise mechanism of action requires further investigations, this drug probably enhances both monoaminergic transmission and GABAergic inhibition. Clinical studies failed to prove any advantage in the emergence time compared to placebo.[17]

In preclinical investigations, caffeine also offered important prospects for accelerating awakening from anesthesia.[18] It is a methylxanthine derivative that works by blocking adenosine receptors which in turn leads to vasoconstriction of cerebral blood vessels. Again, caffeine stimulates sodium–potassium pumps to increase

CSF production and increases sensitivity to catecholamine. Clinically, Fong et al.[19] demonstrated that in healthy individuals ($n = 8$) anesthetized twice with 1.2% isoflurane for 1 hour, the intravenous administration of caffeine citrate (15 mg/kg) during the final 10 minutes of each session, induced a significant ($p = .002$) reduction in awakening times (42%) compared to placebo. Other clinical studies on the topic are ongoing (NCT03577730), although the results have not yet been presented.

Other compounds that could be used are dextroamphetamine and atomoxetine, physostigmine, and nicotine. However, only preclinical evidence is currently available.[16,20,21]

Anesthetics and consciousness: the "awareness" paradox

Anesthetics and GA represent an incredible opportunity to investigate fields that otherwise would be nearly impossible to reach. The study of memory phenomena, and above all the mechanisms of consciousness, finds a "natural" benchside during anesthesia.[22]

Although unconsciousness and amnesia are the main goals and the pillars of GA, and these objectives are achieved almost all patients underwent GA, in very exceptional cases accidental consciousness and memorization may occur. This anesthesia complication is generally termed as GA awareness or accidental awareness during GA or simply "awareness." It represents a fascinating matter of investigation, intersecting numerous study areas such as the pharmacodynamics of anesthetics and the mechanisms of GA, the relationship between anesthesia and consciousness, the processes of memorization, and the application of this knowledge for developing technologies useful for intraoperative brain monitoring.[23]

The awareness phenomenon encompasses two subgroups:

- The anesthesia awareness with recall
- The awareness without explicit recall

They differ from the underlying memory pathway that has been activated. In particular, the awareness with recall requires activation of the declarative (explicit) memory, while the nondeclarative (implicit) memory is involved in the mechanisms of the awareness without conscious content.[24,25]

From the perspective of translational medicine, preclinical research and knowledge from other fields of study have provided a wealth of information that is helping us to solve a puzzle that, until a few years ago, seemed impossible to do. Despite this, many dark sides remain.

These pieces of the puzzle are above all the previously summarized mechanisms of action of anesthetics as well as the neural correlates of memory and consciousness. Since consciousness and memory are separate cognitive processes, anesthetics impact these processes differently. For instance, it was proved that the doses required for unconsciousness are generally higher than those required for amnesia.[26]

The definition of consciousness and related biomolecular mechanisms that determine it would introduce the reader into a minefield. The concept to be emphasized, as it is useful for the narrative path, is that even under total anesthesia, some brain regions are still able to process sensations of the surrounding environment, even if these inputs cannot be recalled upon waking. In other words, the connectivity fragmentation process is partial and dynamic.

Concerning memory, the neural correlates and mechanisms of explicit memory differ from those of implicit memory and anesthetics have different impacts on both. Regardless of the type of memory that is activated and the circuits that work to define it (i.e., working memory), consolidation represents the key process, as it allows the information acquired to be stored for a practically unlimited time. The translational research on memory processing showed that consolidation is not a static and unidirectional process. A stabilized memory can return to a labile state and, in turn, can undergo a subsequent consolidation phase, called reconsolidation.[27] In this view, the recall is an expression of repeated processes of consolidation and reconsolidation and each reconsolidation phase may be altered by several types of interference, including behavioral, psychological, and pharmacological.

By summarizing, memory processing encompasses encoding (i.e., acquiring and representing of memory), storage through the working memory stage, and consolidation. This latter process involves synaptic and broad cellular events such as transcriptional, translational, posttranslational mechanisms, and feedback and feedforward regulation. Consolidation can be followed by retrieving the acquired information. All this work involves brain areas even anatomically distant from each other. In particular, explicit memories ("knowing what"), that include autobiographical memories of events (episodic memory), and general facts and information (semantic memory), are encoded by the hippocampus (especially the episodic memory), entorhinal cortex, and perirhinal cortex although consolidated and stored in the temporal cortex. On the other side, there is the unconscious memory. It represents a set of procedural memories ("knowing how") encompassing the unconscious memory of skills and how to do things and that are encoded by the putamen, caudate nucleus, motor cortex, and cerebellum involved in motor control. The amygdala is implicated in both memory processing mechanisms as it adds emotional contents to explicit and implicit memories.

How does memory work during anesthesia? What is the action of anesthetics on memorization processes? Anesthetics may interfere with working memory and consolidation processes. For instance, propofol may induce anterograde amnesia by interfering with the episodic memory system.[28] Again, although low doses of this anesthetic did not affect working memory, it prevented retention of material in long-term memory.[29] It seems to be mainly a problem of doses as, during surgical anesthesia, propofol also affects working memory. Conflicting results emerged from preclinical studies conducted on other anesthetics. Thiopental and methohexital have probably poor amnesic action[30] while etomidate manifested an important amnesic effect, exerted through $GABA_A$ receptor-mediated long-term potentiation inhibition.[31] Thus, it was postulated that, during etomidate administration, $GABA_A$ receptors are involved in amnesic but not sedative or hypnotic processes.

Concerning ketamine, preclinical research highlighted that its amnesic properties are due to the inhibition of hippocampal $\alpha4\beta2$ nicotinic acetylcholine receptors,[32] which modulates the synaptic release of neurotransmitters in the hippocampus. The NMDA receptors are implicated in memory formation. Moreover, In vivo experiments demonstrated that this agent may induce both anterograde and retrograde amnesia via blocking the NMDA receptors, or by impacting the glycogen synthase kinase 3β/β-catenin signaling pathway. This latter finding has great translational importance as this pathway regulates the transcription of genes involved in processes of synaptic plasticity and memory mechanisms.[33]

Inhalational anesthetics, including volatile anesthetic compounds such as halothane, isoflurane, sevoflurane, and desflurane, as well as certain anesthetic gases, such as nitrous oxide and xenon may prevent memory processing at subhypnotic concentrations. They exert this action mainly through the impairment of the hippocampal θ-rhythm (4–12 Hz) that is the functional correlate of the plasticity activation and memory network processing integration.[34] Xenon exhibits a significant amnesic effect via inhibition of both NMDA and quisqualate receptors in amygdala neurons.[35]

In summary, in this area of study, translational research can work in a bidirectional context from the clinic to the benchside. The clinic can make use of basic notions on the mechanism of action of anesthetics for studying the occurrence of paradoxes such as awareness[36] and, in turn, for testing more effective preventive strategies.[37] However, the most fascinating aspect is the possibility of anesthesia to offer an experimental model useful for understanding cognitive functions. A model that responds to a quantum-type mathematical functional mechanism rather than responding to a simple on/off mechanism.[38] Researchers engaged in the study of the mechanisms of memory and consciousness, and the processes of connectivity, use the means of anesthesia to translate acquisitions in several biomedical fields, such as those related to the investigation on severe brain damage and coma.

General anesthetics and neurodegenerative processes

A large number of studies focused on the potential role of general anesthetics in the progression of neurodegenerative conditions. The main concern regards the potential association between anesthetic exposure and postoperative impairment of the cognitive trajectory. This phenomenon can be clinically expressed as postoperative cognitive dysfunction,[39,40] and progression of neurodegenerative diseases such as Alzheimer's disease (AD). Nevertheless, although it was postulated that inhaled anesthetics may lead to an acceleration of the cognitive decline in AD, this cause-effect role is still debated. The models used to study the pathophysiology of AD have been widely used for research in this area. This already represents an excellent example of translational medicine.

Many experiments on AD transgenic (Tg) mouse models have been carried out for investigating the effects of anesthetics exposure on AD progression.[41] Overall, the results are not conclusive as they depend on the different experimental conditions and the type of general anesthetic tested. Concerning inhaled anesthetics, it was shown that these agents may promote the development of extracellular amyloid plaques.[42] In the context of translational research, it must be clarified because halothane (0.8%–1% in 30% oxygen for 5 days) but not isoflurane (0.9%–1% in 30% oxygen for 5 days) exposure can induce amyloidopathy, in mice. Moreover, when the cognitive performance and behavior were assessed through the Morris Water Maze test, the decline was not found in the halothane-exposed mice but only in animals under isoflurane exposure.

On the other hand, it was demonstrated that younger female mice treated with halothane (0.9%–1.1% corresponding to 1.0 minimal alveolar concentration, MAC) but not isoflurane (1.0 MAC) showed an improvement in memory.[43] Of note, there was a discrepancy between the behavior and pathologic findings as, after 6 months of exposure, degenerative processes were observed in both groups. Again, Su et al.[44] proved that in presymptomatic mice (i.e., during amyloid deposition) isoflurane improved spatial memory and learning, and attenuated the development of the plaques, in the hippocampus. All these discrepancies represent a typical gap that can be bridged by more focused translational medicine approaches.

The amyloid-beta proteins (APP) production and aggregation, and altered cleavage is a complex process of key importance in AD pathogenesis. It was previously demonstrated that inhaled anesthetic may enhance APP oligomerization and cytotoxicity.[45] Animal models of AD were adopted to evaluate the potential of some substances to antagonize the degenerative effects experimentally produced by anesthetics. In APP mutant mice with isoflurane-induced hippocampal and cortical neurodegenerative lesions, the administration of the disaccharide trehalose (1% in fresh drinking water) counteracted significantly the development of these alterations. In particular, this substance reduced the amount of amyloid peptide aggregates and phospho-tau levels. The authors showed that, even if produced, the pro-amyloidogenic effects experimentally provoked by inhaled anesthetics can be effectively prevented.[46]

Another paramount process in the pathogenic cascade of AD is performed by the intraneuronal neurofibrillary tangles. These structures consist of tau proteins—soluble proteins with a pivotal role in maintaining the stability of microtubules in axons—atypically hyperphosphorylated. In genetically modified tau mice, Menuet et al.[47] illustrated that isoflurane accelerated tauopathy, producing, in turn, behavioral changes in presymptomatic animals. Interestingly, these morphological findings were prevented by hyperventilation, or memantine, an NMDA antagonist used in AD therapy. The authors demonstrated that NMDA-receptors are involved in the mechanisms underlying the isoflurane-induced neurodegeneration. Again, the study of strategies useful for counteracting these changes could be translated into clinical studies in AD patients or on those suffering from minimal cognitive impairment.

In another investigation, the authors evaluated the effects of isoflurane administration on pathologic findings suggestive of neurodegeneration and on learning and memory tasks in zinc-deficient mice. Since zinc is essential for learning and memory processes, its deficiency may worsen AD pathology. The hypothesis was demonstrated as mice showed significant cognitive gaps. It was suggested that zinc treatment may enhance memory function and learning in AD predisposed mice and may be useful for preventing neurotoxicity induced by particular substances.[48]

Later on, Lu et al.[49] tested the effects of sevoflurane in newborns Tg mice with APP mutations. While the inhaled agent precipitated the altered APP processing and enhanced brain amyloid fragment (Aβ) levels, the pretreatment with an inositol 1,4,5-trisphosphate receptor antagonist, was able to mitigate these devastating effects.

Different findings were obtained when the intravenous anesthetic propofol was used. Several investigations found that this agent has important neuroprotective features.[50] Results from in vitro and in vivo studies showed that propofol reduced the activation of caspase-3—that contributes to synaptic plasticity and cognitive dysfunction, mostly in the early stage of AD—previously induced by isoflurane, and attenuated the oligomerization of amyloid fractions.[51]

All these data suggest that propofol investigation can offer interesting translational perspectives in the field of neurodegenerative diseases. For example, in vivo and in vitro investigations demonstrated that chronic treatment with propofol (50 mg/kg/week, intraperitoneally for 8–12 weeks) lowered the Aβ-induced mitochondrial dysfunction and improved cognitive function.[52] Again, recent studies highlighted the positive effects of propofol on neuropathology and cognition in AD Tg-mouse models. For instance, Mardini et al.[53] proved that propofol anesthesia was able to produce better cognitive effects in aged mice. Interestingly, neuroinflammation (tested by biochemical tests), and tau aggregation and plaque formation, were unaltered in Tg propofol-treated mice, with or without surgery. This last result offers a further suggestion for translational research, laying the foundations for defining the weight of surgery (stress surgery) and anesthesia in the determinism of postoperative neurocognitive alterations.

Impact on carcinogenic processes

Since anesthetics are compounds with many biological properties, at least experimentally, they can interfere with multiple biological functions. A very interesting aspect that could open many translational research scenarios, concerns the impact of these compounds with carcinogenic processes.[54] The premise is that anesthetics may exert direct modulatory effects on the innate and adaptive immune responses but also numerous indirect actions on immunity, mostly mediated by neuroendocrine responses.[55]

Results obtained from the in vivo and in vivo studies on this hot topic do not allow to arrive at firm conclusions. A substantial part of these studies underlines the protective role of anesthetics. For instance, preclinical studies indicated that, due to its strong antiinflammatory and antioxidative properties, propofol may inhibit cancer recurrence and metastasis. These in vitro experiments were conducted on different cancer cells including pancreatic,[56] colon,[57] breast,[58] hepatocellular carcinoma,[59] ovarian,[60] and prostate cancer cells,[61] and showed that this intravenous agent may slow down the growth of cancer cells by impacting on several signaling pathways.

On the contrary, other researchers have produced diametrically opposite results. In other preclinical studies, it was proved that propofol may promote breast cancer cell progression by modulating different molecular pathways. For instance, Garib et al.[62] demonstrated that the propofol-induced activation of the $GABA_A$ receptor led to an increased migration of breast cancer cells. This effect was produced via calcium influx and reorganization of the actin cytoskeleton.

The contradictory results obtained from the preclinical research could be associated with the heterogeneous experimental conditions presented including concentrations of the anesthetic adopted, cancer cells used, schedules of treatment, and signaling pathways investigated.

Moreover, clinical data on this topic are scarce and derive from retrospective analyses. In one of these analyses, conducted in the setting of breast cancer surgery, the authors found that, compared with volatile anesthesia, propofol improved survival rates and reduced cancer recurrence. Nevertheless, after correcting of confounders, these discrepancies were not statistically significant.[63] In another research on the same outcomes, the effects of propofol-based total intravenous anesthesia (TIVA) were analogous to those of volatile anesthesia.[64] Again, Huang et al.[65] demonstrated that propofol-based TIVA did not improve survival compared to desflurane anesthesia.

On the contrary, in another retrospective analysis, Lee et al.[66] concluded that propofol TIVA reduced the risk of breast cancer recurrence, until five years after surgery. Similar results were obtained by Wingmore et al.[67] who showed that, compared to cancer women receiving volatile anesthetics, cancer patients who underwent propofol anesthesia had significantly better long-term survival rates. In another setting, the authors showed that, compared with desflurane anesthesia, propofol administration was associated with a significant survival improvement in hepatocellular carcinoma patients who underwent hepatectomy ($p < .001$).[68] Moreover, a survival analysis carried out on patients who underwent glioblastoma surgery demonstrated that neither isoflurane nor desflurane affected the survival of cancer patients.[69]

These contrasting results suggest that, in this area, translational research must necessarily be strengthened; preclinical research must provide more indications for clinical research, which in turn must be directed towards well-designed prospective experiments. It is necessary, for example, to evaluate the weight exerted by the different elements of anesthesia, such as general anesthetics and approaches,[70] perioperative opioids,[71–73] and the techniques (e.g., loco-regional/opioid-free strategies) used, on surgery/anesthesia-induced immunosuppression and, in turn, on postoperative oncological outcomes.

Perspectives

In the field of anesthetics and anesthesia, several translational perspectives must be considered. It is mandatory, for example, to refer to large expert-curated databases of relevant documents that exhaustively may cover research fields closer to anesthesiology.[74] This "simplification" process can help researchers to identify the main topics on which their research can be focused. Consequently, this process can allow establishing with which disciplines, precise collaborative programs can be developed.

The use of ad hoc animal models is another goal that anesthetic research must pursue. In this regard, the National Institutes of Health has long supported using nonhuman primate models for research on organ transplantation.[75] It represents a unique opportunity for anesthesia because adopting these models is very difficult due to bureaucratic, organizational, and economic problems. Therefore, collaborative projects could be designed exploiting these "advanced" models for anesthesia research purposes.

An important prospect for the enhancement of translational research skills in the anesthesiology field must be the improvement of training areas. There is an obvious difficulty in disseminating scientific findings, especially when new acquisition can potentially involve modifications of the clinical practice. How long did it take to standardize the use of neuromuscular monitoring in the theater, despite scientific evidence have significantly emphasized the great advantages in terms of outcomes?

Finally, the issue of anesthetics development. This process must necessarily be the effect of translational research. In this context, artificial intelligence can play a very important role. By referring to databases containing the characteristics of infinite synthesized molecules, it could be possible to intercept potential anesthetics that act on particular receptors, and manifest favorable pharmacokinetic profiles (e.g., very rapid elimination), and low toxicity. This in silico strategy could also drive the synthesis of new molecules. On these bases, Cayla et al.[76] developed a highly selective $GABA_A$-slow receptor agent (i.e., compound BB) with anesthetic properties and minimal hemodynamic side effects. Later on, they tested it on rodent brain slices and Xenopus oocytes, and in vivo in rats for evaluating hemodynamic effects.

Conclusion

Because anesthetic drugs have multiple biological properties, they can potentially interfere with multiple processes. For this reason, anesthetics can ideally be placed at the center of research programs which, from the preclinical phase to the clinical phase, involve several biomedical areas, and beyond. Research fields such as computer science, artificial intelligence, bioengineering, can find incredible fertile ground in this sector. The most effective strategy is to train professionals who, by deciphering a Rosetta Stone, can interpret the complex language of the individual disciplines involved, directing research towards the set objectives, or designing new objectives starting from what, basic or clinical research, proposes.

References

1. Cohrs Jr R, Martin T, Ghahramani P, Bidaut L, Higgins Jr P, Shahzad A. Translational medicine definition by the European Society for Translational Medicine. *N Horiz Transl Med.* 2014;2(3):86–88.

2. Butler D. Translational research: crossing the valley of death. *Nature*. 2008;453: 840–842.
3. Fernández L, Velásquez M, Sua LF, et al. The porcine biomodel in translational medical research: from biomodel to human lung transplantation. *Biomedica*. 2019;39(2): 300–313.
4. Cahill TJ, Choudhury RP, Riley PR. Heart regeneration and repair after myocardial infarction: translational opportunities for novel therapeutics. *Nat Rev Drug Discov*. 2017;16(10):699–717.
5. Khan RMM, Chua ZJY, Tan JC, Yang Y, Liao Z, Zhao Y. From pre-diabetes to diabetes: diagnosis, treatments and translational research. *Medicina (Kaunas)*. 2019;55(9):546.
6. Cascella M, Bimonte S, Muzio MR, Schiavone V, Cuomo A. The efficacy of epigallocatechin-3-gallate (green tea) in the treatment of Alzheimer's disease: an overview of pre-clinical studies and translational perspectives in clinical practice. *Infect Agent Cancer*. 2017;12:36. Available from: https://doi.org/10.1186/s13027-017-0145-6.
7. Bimonte S, Albino V, Piccirillo M, et al. Epigallocatechin-3-gallate in the prevention and treatment of hepatocellular carcinoma: experimental findings and translational perspectives. *Drug Des Dev Ther*. 2019;13:611–621. Available from: https://doi.org/10.2147/DDDT.S180079.
8. Cuomo A, Bimonte S, Forte CA, Botti G, Cascella M. Multimodal approaches and tailored therapies for pain management: the trolley analgesic model. *J Pain Res*. 2019;12:711–714.
9. Buonaguro FM, Botti G, Ascierto PA, et al. The clinical and translational research activities at the INT - IRCCS "Fondazione Pascale" cancer center (Naples, Italy) during the COVID-19 pandemic. *Infect Agent Cancer*. 2020;15(1):69. Available from: https://doi.org/10.1186/s13027-020-00330-7.
10. Imas OA, Ropella KM, Ward BD, Wood JD, Hudetz AG. Volatile anesthetics disrupt frontal-posterior recurrent information transfer at gamma frequencies in rat. *Neurosci Lett*. 2005;387:145–150.
11. Akeju O, Loggia ML, Catana C, et al. Disruption of thalamic functional connectivity is a neural correlate of dexmedetomidine-induced unconsciousness. *Elife*. 2014;3:e04499. Available from: https://doi.org/10.7554/eLife.04499.
12. Cascella M. Emergence from anesthesia: a winding way back. *Anaesthesiol Intensive Ther*. 2018;50(2):168–169.
13. Scheib CM. Brainstem influence on thalamocortical oscillations during anesthesia emergence. *Front Syst Neurosci*. 2017;11:66. Available from: https://doi.org/10.3389/fnsys.2017.00066.
14. de Lecea L, Huerta HR. Hypocretin (orexin) regulation of sleep-to-wake transitions. *Front Pharmacol*. 2014;5:16. Available from: https://doi.org/10.3389/fphar.2014.00016.
15. Yang C, Zhang L, Hao H, Ran M, Li J, Dong H. Serotonergic neurons in the dorsal raphe nucleus mediate the arousal-promoting effect of orexin during isoflurane anesthesia in male rats. *Neuropeptides*. 2019;75:25–33. Available from: https://doi.org/10.1016/j.npep.2019.03.004.
16. Kenny JD, Chemali JJ, Cotten JF, et al. Physostigmine and methylphenidate induce distinct arousal states during isoflurane general anesthesia in rats. *Anesth Analg*. 2016;123(5):1210–1219.
17. Xará D, Silva A, Mendonça J, Abelha F. Inadequate emergence after anesthesia: emergence delirium and hypoactive emergence in the Postanesthesia Care Unit. *J Clin Anesth*. 2013;25(6):439–446. Available from: https://doi.org/10.1016/j.jclinane.2013.02.011.

18. Wang Q, Fong R, Mason P, Fox AP, Xie Z. Caffeine accelerates recovery from general anesthesia. *J Neurophysiol*. 2014;111:1331–1340. Available from: https://doi.org/10.1152/jn.00792.2013.
19. Fong R, Wang L, Zacny JP, et al. Caffeine accelerates emergence from isoflurane anesthesia in humans: a randomized, double-blind, crossover study. *Anesthesiology*. 2018;129:912–920.
20. Kenny JD, Taylor NE, Brown EN, Solt K. Dextroamphetamine (but not atomoxetine) induces reanimation from general anesthesia: implications for the roles of dopamine and norepinephrine in active emergence. *PLoS One*. 2015;10(7):e0131914.
21. Alkire MT, McReynolds JR, Hahn EL, et al. Thalamic microinjection of nicotine reverses sevoflurane-induced loss of righting reflex in the rat. *Anesthesiology*. 2007;107:264–272.
22. Cascella M. The challenge of accidental awareness during general anesthesia. In: Cascella M, ed. *General Anesthesia Research. Neuromethods*. Vol 150. United States:: Springer; 2020:1–32.
23. Cascella M. Mechanisms underlying brain monitoring during anesthesia: limitations, possible improvements, and perspectives. *Korean J Anesthesiol*. 2016;69(2):113–120.
24. Cascella M. What about memory, consciousness, recall, and awareness in anesthesia? *Iran J Med Sci*. 2014;39(3):311–312.
25. Cascella M, Viscardi D, Schiavone V, et al. A 7-year retrospective multisource analysis on the incidence of anesthesia awareness with recall in cancer patients: a chance of collaboration between anesthesiologists and psycho-oncologists for awareness detection. *Med (Baltim)*. 2016;95(5):e2757. Available from: https://doi.org/10.1097/MD.0000000000002757.
26. Veselis RA, Reinsel RA, Feshchenko VA, Johnson Jr. R. Information loss over time defines the memory defect of propofol: a comparative response with thiopental and dexmedetomidine. *Anesthesiology*. 2004;101(4):831–841.
27. Takamiya S, Yuki S, Hirokawa J, Manabe H, Sakurai Y. Dynamics of memory engrams. *Neurosci Res*. 2019;153:22–26. Available from: https://doi.org/10.1016/j.neures.2019.03.005. pii: S0168–0102(19) 30089–6.
28. Tulving E. Episodic memory and common sense: how far apart? *Philos Trans R Soc Lond Ser B Biol Sci*. 2001;356(1413):1505–1515.
29. Veselis RA, Pryor KO, Reinsel RA, Li Y, Mehta M, Johnson Jr. R. Propofol and midazolam inhibit conscious memory processes very soon after encoding: an event related potential study of familiarity and recollection in volunteers. *Anesthesiology*. 2009;110(2):295–312.
30. Veselis RA, Reinsel RA, Feshchenko VA, Wroń ski M. The comparative amnestic effects of midazolam, propofol, thiopental, and fentanyl at equisedative concentrations. *Anesthesiology*. 1997;87:749–764.
31. Zarnowska ED, Rodgers FC, Oh I, et al. Etomidate blocks LTP and impairs learning but does not enhance tonic inhibition in mice carrying the N265M point mutation in the beta3 subunit of the GABA (A) receptor. *Neuropharmacology*. 2015;93:171–178.
32. Forman SA, Chin VA. General anesthetics and molecular mechanisms of unconsciousness. *Int Anesthesiol Clin*. 2008;46:43–53.
33. Arrázola MS, Varela-Nallar L, Colombres M, et al. Calcium/calmodulin-dependent protein kinase type IV is a target gene of the wnt/beta-catenin signaling pathway. *J Cell Physiol*. 2009;221:658–667.
34. Masquelier T, Hugues E, Deco G, Thorpe SJ. Oscillations, phase-of-firing coding, and spike timing-dependent plasticity: an efficient learning scheme. *J Neurosci*. 2009;29(43):13484–13493.

35. Haseneder R, Kratzer S, Kochs E, et al. Xenon reduces N-methyl-D-aspartate and alpha-amino-3-hydroxy-5-methyl-4-isoxazolepropionic acid receptor-mediated synaptic transmission in the amygdala. *Anesthesiology*. 2008;109:998–1006.
36. Cascella M, Bimonte S, Amruthraj NJ. Awareness during emergence from anesthesia: Features and future research directions. *World J Clin Cases*. 2020;8(2):245–254.
37. Cascella M. Anesthesia awareness. Can midazolam attenuate or prevent memory consolidation on intraoperative awakening during general anesthesia without increasing the risk of postoperative delirium? *Korean J Anesthesiol*. 2015;68(2):200–202. Available from: https://doi.org/10.4097/kjae.2015.68.2.200.
38. Cascella M, Schiavone V, Muzio MR, Cuomo A. Consciousness fluctuation during general anesthesia: a theoretical approach to anesthesia awareness and memory modulation. *Curr Med Res Opin*. 2016;32(8):1351–1359.
39. Cascella M, Bimonte S. The role of general anesthetics and the mechanisms of hippocampal and extra-hippocampal dysfunctions in the genesis of postoperative cognitive dysfunction. *Neural Regen Res*. 2017;12(11):1780–1785.
40. Cascella M, Muzio MR, Bimonte S, Cuomo A, Jakobsson JG. Postoperative delirium and postoperative cognitive dysfunction: updates in pathophysiology, potential translational approaches to clinical practice and further research perspectives. *Minerva Anestesiol*. 2018;84(2):246–260.
41. Quiroga C, Chaparro RE, Karlnoski R, et al. Effects of repetitive exposure to anesthetics and analgesics in the Tg2576 mouse Alzheimer's model. *Neurotox Res*. 2014;26(4):414–421.
42. Eckenhoff RG, Johansson JS, Wei H, et al. Inhaled anesthetic enhancement of amyloid-beta oligomerization and cytotoxicity. *Anesthesiology*. 2012;101(3):703–709.
43. Tang JX, Mardini F, Caltagarone BM, et al. Anesthesia in presymptomatic Alzheimer's disease: a study using the tripletransgenic mouse model. *Alzheimers Dement*. 2011;7(5):521–531.
44. Su D, Zhao Y, Xu H, et al. Isoflurane exposure during mid-adulthood attenuates age-related spatial memory impairment in APP/PS1 transgenic mice. *PLoS One*. 2012;7(11):e50172.
45. Eckenhoff G, Johansson JS, Wei H, et al. Inhaled anesthetic enhancement of amyloid beta oligomerization and cytotoxicity. *Anesthesiology*. 2004;101:703–709.
46. Perucho J, Casarejos MJ, Gomez A, et al. Trehalose protects from aggravation of amyloid pathology induced by isoflurane anesthesia in APP(swe) mutant mice. *Curr Alzheimer Res*. 2012;9(3):334–343.
47. Menuet C, Borghgraef P, Voituron N, et al. Isoflurane anesthesia precipitates tauopathy and upper airways dysfunction in pre-symptomatic Tau. P301L mice: possible implication for neurodegenerative diseases. *Neurobiol Dis*. 2012;46(1):234–243.
48. Feng C, Liu Y, Yuan Y, et al. Isoflurane anesthesia exacerbates learning and memory impairment in zinc-deficient APP/PS1 transgenic mice. *Neuropharmacology*. 2016;111:119–129.
49. Lu Y, Wu X, Dong Y, et al. Anesthetic sevoflurane causes neurotoxicity differently in neonatal naive and Alzheimer disease transgenic mice. *Anesthesiology*. 2010;112:1404–1416.
50. Bayona NA, Gelb AW, Jiang Z, et al. Propofol neuroprotection in cerebral ischemia and its effects on low-molecular-weight antioxidants and skilled motor tasks. *Anesthesiology*. 2004;100:1151–1159.
51. Zhang Y, Zhen Y, Dong Y, et al. Anesthetic propofol attenuates the isoflurane-induced caspase-3 activation and Aβ oligomerization. *PLoS One*. 2011;6(11):e27019.

52. Shao H, Zhang Y, Dong Y, et al. Chronic treatment with anesthetic propofol improves cognitive function and attenuates caspase activation in both aged and Alzheimer's disease transgenic mice. *J Alzheimers Dis*. 2014;41(2):499–513.
53. Mardini F, Tang JX, Li JC, et al. Effects of propofol and surgery on neuropathology and cognition in the 3xTgAD Alzheimer transgenic mouse model. *Br J Anaesth*. 2017; 119(3):472–480.
54. Bimonte S, Cascella M, Giudice A, et al. *Propofol effects in breast cancer cell progression: evidences from in vitro studies. Marco C. ed. General Anesthesia Research. Neuromethods*. Vol 150. United States:: Springer; 2020:147–157.
55. Stollings LM, Jia LJ, Tang P, Dou H, Lu B, Xu Y. Immune Modulation by Volatile Anesthetics. *Anesthesiology*. 2016;125(2):399–411.
56. Liu ZM, Zhang J, Hong GC, et al. Propofol inhibits growth and invasion of pancreatic cancer cells through regulation of the miR-21/Slug signaling pathway. *Am J Transl Res*. 2016;8:4120–4133.
57. Miao Y, Zhang Y, Wan H. GABA-receptor agonist, propofol inhibits invasion of colon carcinoma cells. *Biomed Pharmacother*. 2010;64(9):583–588. Available from: https://doi.org/10.1016/j.biopha.2010.03.006.
58. Ecimovic P, Murray D, Doran P, Buggy DJ. Propofol and bupivacaine in breast cancer cell function in vitro - role of the NET1 gene. *Anticancer Res*. 2014;34(3): 1321–1331.
59. Du W, Lv J, Zou X, et al. Propofol inhibits hepatocellular carcinoma growth and invasion through the HMGA2 - mediated Wnt/beta - catenin pathway. *Exp Ther Med*. 2017;13(5):2501–2506. Available from: https://doi.org/10.3892/etm.2017.4253.
60. Wang P, Chen J, Mu LH, et al. Propofol inhibits invasion and enhances paclitaxel-induced apoptosis in ovarian cancer cells through the suppression of the transcription factor slug. *Eur Rev Med Pharmaeol Sei*. 2013;17:1722–1729.
61. Qian J, Shen S, Chen W, et al. Propofol reversed hypoxia-induced docetaxel resistance in prostate cancer cells by preventing epithelial-mesenchymal transition by inhibiting hypoxia-inducible factor la. *Biomed Res Int*. 2018;2018:4174232. Available from: https://doi.org/10.1155/2018/4174232.
62. Garib V, Lang K, Niggemann B, Zänker KS, Brandt L, Dittmar T. Propofol-induced calcium signalling and actin reorganization within breast carcinoma cells. *Eur J Anaesthesiol*. 2005;22(8):609–615. Available from: https://doi.org/10.1017/s026502150500102x.
63. Enlund M, Berglund A, Andreasson, et al. The choice of anaesthetic–sevoflurane or propofol–and outcome from cancer surgery: a retrospective analysis. *Ups J Med Sci*. 2014;119(3):251–261.
64. Kim MH, Kim DW, Kim JH, Lee KY, Park S, Yoo YC. Does the type of anesthesia really affect the recurrence-free survival after breast cancer surgery? *Oncotarget*. 2017;8 (52):90477–90487.
65. Huang YH, Lee MS, Lou YS, et al. Propofol-based total intravenous anesthesia did not improve survival compared to desflurane anesthesia in breast cancer surgery. *PLoS One*. 2019;14(11):e0224728.
66. Lee JH, Kang SH, Kim Y, et al. Effects of propofol-based total intravenous anesthesia on recurrence and overall survival in patients after modified radical mastectomy: a retrospective study. *Korean J Anesthesiol*. 2016;69(2):126–132.
67. Wigmore TJ, Mohammed K, Jhanji S. Long-term survival for patients undergoing volatile vs iv anesthesia for cancer surgery: a retrospective analysis. *Anesthesiology*. 2016;124(1):69–79. Available from: https://doi.org/10.1097/ALN.0000000000000936.

68. Lai HC, Lee MS, Lin C, et al. Propofol-based total intravenous anaesthesia is associated with better survival than desflurane anaesthesia in hepatectomy for hepatocellular carcinoma: a retrospective cohort study. *Br J Anaesth.* 2019;123(2):151–160.
69. Cata JP, Hagan KB, Bhavsar SD, et al. The use of isoflurane and desflurane as inhalational agents for glioblastoma surgery. A survival analysis. *J Clin Neurosci.* 2017; 35:82–87.
70. Cascella M, Quarto G, Grimaldi G, et al. Neuropathic painful complications due to endopelvic nerve lesions after robot-assisted laparoscopic prostatectomy: Three case reports. *Med (Baltim).* 2019;98(46):e18011. Available from: https://doi.org/10.1097/MD.0000000000018011.
71. Bimonte S, Barbieri A, Cascella M, et al. Naloxone counteracts the promoting tumor growth effects induced by morphine in an animal model of triple-negative breast cancer. *Vivo.* 2019;33(3):821–825. Available from: https://doi.org/10.21873/invivo.11545.
72. Bimonte S, Barbieri A, Cascella M, et al. The effects of naloxone on human breast cancer progression: in vitro and in vivo studies on MDA. MB231 cells. *Onco Targets Ther.* 2018;11:185–191. Available from: https://doi.org/10.2147/OTT.S145780.
73. Bimonte S, Cascella M, Barbieri A, Arra C, Cuomo A. Shining a light on the effects of the combination of (-)-epigallocatechin-3-gallate and tapentadol on the growth of human triple-negative breast cancer cells. *Vivo.* 2019;33(5):1463–1468.
74. Brown P, RELISH Consortium, Zhou Y. Large expert-curated database for benchmarking document similarity detection in biomedical literature search. *Database (Oxf).* 2019;2019:baz085. Available from: https://doi.org/10.1093/database/baz085.
75. Knechtle SJ, Shaw JM, Hering BJ, Kraemer K, Madsen JC. Translational impact of NIH-funded nonhuman primate research in transplantation. *Sci Transl Med.* 2019; 11(500):eaau0143. Available from: https://doi.org/10.1126/scitranslmed.aau0143.
76. Cayla NS, Dagne BA, Wu Y, et al. A newly developed anesthetic based on a unique chemical core. *Proc Natl Acad Sci U S A.* 2019;116(31):15706–15715.

Neurotoxicity and translational research

4

Zachary Fleissner[1], Antonia Schonwald[2], Neil Patel[1], Benjamin F. Gruenbaum[1], Valerie A. Smart[3] and Shaun E. Gruenbaum[1]
[1]Department of Anesthesiology and Perioperative Medicine, Mayo Clinic-Florida, Jacksonville, FL, United States, [2]New York Medical College, Valhalla, NY, United States, [3]Department of Aviation Medicine, Naval Hospital Jacksonville, Jacksonville, FL, United States

Introduction

Approximately 40 million anesthetics are delivered annually in the United States, and more than 90% of these anesthetics are administered either directly or under the supervision of an Anesthesiologist. After surgery and anesthesia, a subset of patients might experience a decline in cognitive function that can persist for weeks or even months after surgery. Studying this condition, termed "postoperative cognitive decline" (POCD), is challenging due to varying definitions of POCD between studies, ethical challenges related to randomization and the use a reliable "control" group, and a lack of consensus for which testing modalities and phenotypes best reflect POCD.[1] Although the true incidence of POCD is difficult to ascertain and varies depending on the definitions used and populations studied, the current literature suggests that underlying factors that contribute to the development of POCD are likely multifactorial, and POCD can occur after minor or major surgery.[1]

Exposure to neurotoxic agents in the perioperative period might contribute to the development of POCD, and translational research provides the ideal investigative platform to study these effects as well as the underlying mechanisms and potential neuroprotective strategies. Over the past 20 years, several preclinical (i.e., laboratory) studies have established a potential causative link between prolonged or repeated exposure to commonly used volatile and intravenous anesthetics and neurotoxicity in animal models. These effects have been especially prominent in both extremes of age, which has prompted a public health concern and significant media exposure. As the body of preclinical evidence grew and the underlying mechanisms of anesthesia-induced neurotoxicity were explored, anesthesiologists, surgeons, and patients expressed similar concerns and questions about the role of anesthesia exposure and neurocognitive decline.

This concern prompted the Food and Drug Administration (FDA) to issue an advisory statement in 2015 which recommended that nonurgent surgeries in children under 3 years old and exposure to anesthesia for a duration exceeding 3 h

Perioperative Neuroscience. DOI: https://doi.org/10.1016/B978-0-323-91003-3.00016-7

should be avoided whenever possible. However, the vast majority of procedures in those younger than 3 years of age are urgent and emergent, and a fear of anesthesia exposure might result in delaying or canceling procedures that are medically necessary. The preclinical studies and FDA advisory highlighted the need for high-quality prospective clinical studies that address neurotoxicity and strategies of neuroprotection in the perioperative setting. Despite the plethora of preclinical evidence and subsequent FDA warning, clinical studies have failed to reliably demonstrate neurotoxic effects of anesthetics.

In this chapter, we review the various animal models that have been used to study anesthesia-induced neurotoxicity and its potential underlying mechanisms, and explore specific pharmacologic agents, that have been studied, that might provide neuroprotection or prevent anesthesia-induced neurotoxicity in the perioperative setting. We furthermore review the recent clinical studies that have investigated these potential effects in humans. Finally, we examine future models by which this topic can be explored, including the use of biomarkers neuroimaging techniques.

Preclinical evidence and the use of animal models

The use of animal models has been critical to advancing our understanding of anesthesia exposure and its potential neurotoxic effects. To date, several animal models have been used to investigate anesthesia-induced neurotoxicity, ranging from rodent models to nonhuman primates (Table 4.1). Many commonly used anesthetic agents exert primarily either N-methyl-D-aspartate (NMDA) glutamate receptor blocking effects or γ-aminobutyric acid type A ($GABA_A$) receptor-enhancing effects,[2,3] which has been suggested to play an important role in anesthesia-induced neurotoxicity.

Volatile anesthetics

Several preclinical studies have suggested that volatile anesthetics have neurotoxic effects in the developing brain. In one such study, Jevtovic-Todorovic et al. postulated that prolonged early exposure to anesthetic agents might results in neuronal apoptosis in a rodent model.[4] The investigators induced general anesthesia in 7-day-old Sprague Dawley rats with midazolam, nitrous oxide, and isoflurane, and anesthesia was maintained with the same anesthetic regimen for 6 h. Rats in the control group were maintained on a mock (i.e., placebo) anesthetic infusion for 6 h. The investigators used caspase-3 to identify neuronal apoptosis via histopathological analysis followed by optical dissector and fractionation,[5] and found that exposure to the three anesthetic agents resulted in extensive neurodegeneration by inducing neuronal apoptosis. However, rats exposed to nitrous oxide or midazolam alone did not demonstrate an increase in apoptotic markers. Further evidence in rodent models demonstrated that the concentration and length of volatile anesthesia

Table 4.1 Preclinical studies.

Authors, year	Study model	Primary findings
Jevtovic-Todorovic et al. (2003)[4]	Rat	Prolonged early exposure to anesthetic agents leads to neuronal apoptosis
Lu et al.[6] (2016); Xiao (2016)[7]	Mouse	Concentration and length of volatile anesthesia exposure correlates with extent of neurotoxic effects
Lu et al. (2016)[6]; Tao et al. (2016)[8]	Mouse	Isoflurane shows greater neurodegeneration than sevoflurane and desflurane
Brambrink et al. (2010)[9]	Nonhuman primate	Isoflurane exposure showed widespread neuronal apoptosis
Huang et al. (2016a)[10]	Rats	Ketamine attenuates development of hippocampal dentate gyri
Zheng (2015)[11]	Neonatal mice	Upregulated tumor necrosis factor (TNF)-alpha was neurodegenerative, downregulation of TNF-alpha was neuroprotective
Huang et al. (2016b)[12]	Mouse neural stem cells	Propofol impairs stem cell proliferation and inhibits neurogenesis
Milanovic et al. (2016)[13]	Neonatal rats	Propofol anesthesia induced Fas/FasL, mediating apoptotic and inflammation pathways

Notes: Sample of several of the preclinical studies that have investigated the potential neurotoxic effects of anesthetic agents (including the corresponding author, year, type of study, and primary outcomes).

exposure correlates with the extent of its neurotoxic effects,[6,7] and isoflurane exposure was associated with a higher degree of neurodegeneration in the developing mouse brain compared with exposure to sevoflurane[6] or desflurane.[8]

More recently, translational research studies on anesthesia-induced neurotoxicity have extended to include nonhuman primate models, whose brains are thought to more closely resemble the human brain compared with the rodent brain. Brambrink et al. studied the effects of neonatal exposure to the halogenated volatile anesthetic and $GABA_A$ potentiator isoflurane in rhesus macaques. Macaques in the experimental group were exposed either to isoflurane in concentrations that ranged from 0.7% to 1.3% for 5 h, and macaques in the control group were exposed to room air supplemental oxygenation for 5 h without any other intervention.[14] The investigators used anticaspase-3 antibodies to identify evidence of apoptosis on histopathological analysis and found a 13-fold increase in widespread apoptotic cell density throughout the cerebral cortex after 3 h of isoflurane exposure.

Ketamine

The potential neurotoxic effects of ketamine, an NMDA receptor antagonist and commonly used anesthetic, have been extensively studied in animal models.

A recent metaanalysis that reviewed the preclinical literature suggested that ketamine plays an important role in mediating neuroapoptosis in the developing brain.[15] A recent study in a 7-day-old rat model further established ketamine's role in attenuating the development of hippocampal dentate gyri.[10] In a neonatal (2-month-old) mouse study, in vitro analysis of organotypic cultures of hippocampal slices with western blotting analysis was used to establish the role of tumor necrosis factor (TNF)-alpha in regulating ketamine-induced neurotoxicity.[11] The neurotoxic effects observed on histological analysis were associated with several behavioral deficits including impairments in spatial reference and memory tasks, as demonstrated by the use of a water maze test. The deficits were particularly pronounced in mice with an upregulation of TNF-alpha compared to those with downregulated TNF-alpha function. Furthermore, mice with upregulated TNF-alpha demonstrated increased hippocampal apoptosis. Taken together, the evidence establishes an important link between ketamine exposure, TNF-alpha function, hippocampal apoptosis, and neurocognitive outcomes, and identifies a potential target (TNF-alpha) by which ketamine-induced neurotoxicity might be attenuated.

Propofol

Propofol, a $GABA_A$ potentiator, is one of the most common intravenous anesthetics used in the United States. Its potential neurotoxic effects have been extensively studied, and a growing body of evidence in animal cell culture, animal models (rodents, zebrafish, piglets, primates), and human stem cell-derived neuron cell culture models have demonstrated that early propofol exposure induces neuronal death and potentially lead to developmental delays, learning difficulties, and neurocognitive deficits that persist later in life.[16] Furthermore, several potential mechanisms of propofol-induced neurotoxicity have been suggested and studied. For example, propofol-mediated oligodendrocytic apoptosis was demonstrated to result from Fas receptor induction and Fas ligand expression,[12,13] which thereby interrupts and dysregulates many neural developmental processes. These translational studies provide important insight into the potential mechanisms of propofol-induced neurotoxicity, and suggest that $GABA_A$ receptor might be an important target of novel therapeutic agents that aim to reduce or prevent propofol-induced neurotoxicity.

Anesthesia and neurotoxicity: mechanisms

As highlighted above, various mechanisms that might account for anesthesia-induced neurotoxicity have been proposed, which are thought to be closely related to the underlying pharmacological properties of the anesthetics (Fig. 4.1). Due to their high lipid-solubility, general anesthetics easily penetrate cell membranes and interact with intracellular organelles. These hydrophobic properties contribute to a favorable pharmacokinetic and pharmacodynamic profile, with a fast onset of action and full reversibility. However, the same characteristics allow for a potential

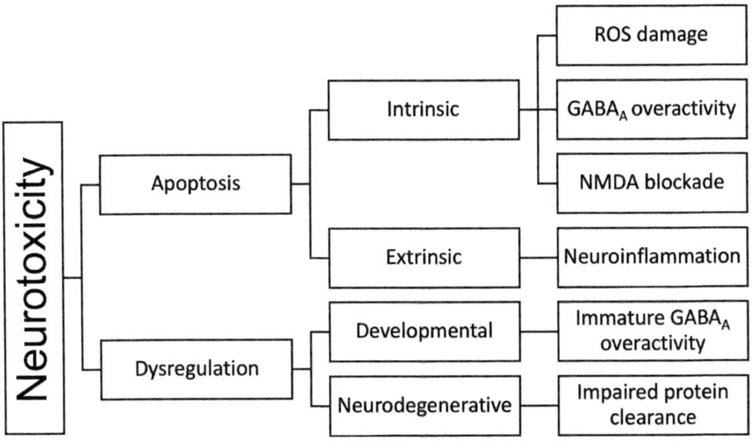

Figure 4.1 Mechanisms of anesthesia-induced neurotoxicity can be divided into apoptosis and dysregulation of cell homeostasis. Apoptosis can occur through the intrinsic or extrinsic pathway. The intrinsic pathway can be triggered by anesthesia-generated reactive oxygen species (ROS) that causes membrane peroxidation, γ-aminobutyric acid type A (GABA$_A$) overactivity that leads to an electrolyte imbalance, or an N-methyl-D-aspartate (NMDA) blockade which causes organelle swelling. The extrinsic pathway is triggered by neuroinflammation which modulates inflammatory factors and the complement cascade. Dysregulation of cell homeostasis is divided into two categories based on the time of insult: developmental and neurodegenerative. Developmental dysregulation is linked to immature GABA$_A$ receptor overactivity in developing brains. Neurodegenerative dysregulation, caused by impaired protein clearance, results in accumulation of deposits and aggregates that potentially contribute to neurodegenerative diseases.

disruption of cellular homeostasis. For instance, preclinical trials note that anesthetics modify gene expression, affect protein synthesis, and cause intracellular stress.[17]

Other long-term consequences also seem to be linked to the impairment of neuronal development and synaptogenesis.[18] The proposed mechanisms for such pathologies include cell-stress-induced apoptosis, and dysregulation of various developmental, growth, and survival pathways.

Anesthesia-induced neuronal apoptosis appears to be further be linked to the activation of both the intrinsic and extrinsic apoptosis pathways. For example, reactive oxygen species (ROS) produced by anesthetics such as ketamine, isoflurane, and nitric oxide (NO) are noted to induce mitochondrial release of cytochrome c which ultimately induces intrinsic apoptosis.[19] This pathway is further potentiated through the activation of GABA$_A$ receptors which increase intracellular calcium and contribute to an increased mitochondrial membrane permeability.[20] NO is specifically noted to also block NMDA receptors, which induces further swelling of other neuronal organelles.[18] Such intracellular stressors, combined with ROS-induced lipid peroxidation of cell membranes, are believed to be a leading cause of cognitive decline seen in preclinical studies. However, these stressors are also often accompanied with neuroinflammation, which is primarily associated with the extrinsic apoptosis pathway.

Interestingly, although clinical studies have demonstrated antiinflammatory properties of several anesthetic agents in humans, animal models have demonstrated an increase in inflammatory mediators with prolonged exposure to anesthesia.[21] Studies have observed an increase in interleukin-6 and TNF-alpha, which act as key mediators of neuroinflammation and might possibly contribute to wide-spread inflammation seen after surgeries.[18] Isoflurane has also been linked to the activation of the complement cascade through the modulation of C1q + and C3 levels,[22] suggesting further complexity of mechanisms underlying neuroinflammation and ultimately apoptosis.

However, anesthesia-associated neurotoxicity is not limited to neuronal cell death, but rather extends to dysregulation of neuronal development and function. Different brain regions are more vulnerable to anesthesia-induced dysregulation at different stages of development and depend upon the degree of exposure. Growing evidence indicates that the two extremes in age seem to be most affected.[18] Susceptibility to anesthetic neurotoxicity at a young age is believed to be related to the changing function of the $GABA_A$ receptor.[17] $GABA_A$ is a major target of many general anesthetics and induces inhibitory and hyperpolarizing effects in the adult brain. However, in the young developing brain, $GABA_A$ exerts different effects and has an excitatory and depolarizing function. Overexcitation neurotoxicity through the stimulation of $GABA_A$ is a plausible explanation for increased susceptibility to neuronal dysregulation at a young age. Such neurotoxicity could potentially leads to impaired neurogenesis and synaptogenesis, which are both critical for the developing brain. At an old age, on the other hand, the brain is more prone to neurodegeneration. Even though the mechanism of many neurodegenerative disorders is still heavily debated, impaired protein clearance seems to be the leading correlate to pathogenesis.

Anesthetics have been observed to contribute to certain neurodegenerative pathologies in preclinical studies. Alzheimer's disease (AD), which is associated with increased deposits of insoluble Aβ protein, can be further potentiated with the use of isoflurane.[23] Most anesthetics also increase hyperphosphorylation of the tau protein[24] which contributes to extracellular tau aggregation also seen in AD. Although these preclinical findings suggest potential causes of neurotoxicity, it is important to note that further investigation clinical trials are necessary to validate whether these effects and mechanisms translate to humans. Interestingly, the limited clinical evidence that is available is less clear, although the studies are limited by several important factors that will be further discussed.

From laboratory to clinical trials: evidence in humans

In contrast to the plethora of preclinical studies in rat, mice, and rhesus macaque models that have demonstrated neurotoxic effects of anesthetics, clinical studies in humans have yielded inconsistent findings. Most of the current clinical data were derived from observational studies, which have demonstrated significant heterogeneity in their measured outcomes (Table 4.2).

Three recent Canadian studies failed to demonstrate an association between repeated and/or prolonged exposures to anesthetic agents early in young children and the later development of neurocognitive deficits as assessed by early development index scores.[25–27] In the Pediatric Anesthesia Neurodevelopment Assessment trial, a matched-sibling observational study included 105 sibling pairs in which one sibling was exposed to anesthesia for inguinal hernia repair before 3 years of age and the other sibling did not have exposure to anesthesia prior to 3 years of age. This study found no differences with respect to intelligence quotient (IQ) scores, neurocognitive function, and subjective behavior assessment between the groups. Similarly, the Mayo Anesthesia Safety in Kids (MASK) trial was a matched cohort

Table 4.2 Clinical studies.

Authors, year	Study design	Primary findings
O'Leary JD et al. (2016)[26], Graham Mr (2016)	Retrospective study: general anesthesia (GA) exposure < 4 years old	Small but significant decrease in early developmental vulnerability scores
O'leary JD (2019)[27]	Retrospective sibling-matched cohort: < 6 years GA exposure versus no GA exposure	No differences in early developmental vulnerability or early developmental index
Sun LS et al. (2016)[28] (PANDA trial)	Sibling matched cohort: 1 GA exposure versus 0 < 36 months	No difference in intelligence quotient scores, neurocognitive function, or behavior assessment
Warner DO et al. (2018)[29] (MASK trial)	Matched cohort study: 0 anesthetics versus 1 anesthetic versus multiple anesthetics in children <36 months	1 anesthetic and multiple anesthetics groups experienced more difficulty with executive function; multiple anesthetics group showed more behavioral problems compared to 0 anesthetics
Walkden GJ et al. (2020)[30] (ASLPAC)	Matched cohort study: 0 GA versus 1 GA versus multiple GA in children < 48 months	No associated clinical or statistically significant neurodegenerative effects
McCann ME et al. (2019)[31] (GAS trial)	Randomized control trial: infants < 60 weeks receiving awake regional anesthesia versus sevoflurane GA for inguinal hernia repair	No difference in neurocognitive deficits at 5 years of age between groups

Notes: Sample of several of the clinical studies that have investigated the potential neurotoxic effects of anesthetic agents (including the corresponding author, year, type of study, and primary outcomes). *ASLPAC*, Avon Longitudinal Study of Parents and Children; *GAS*, general anesthesia compared to spinal anesthesia; *MASK*, Mayo Anesthesia Safety in Kids; *PANDA*, Pediatric Anesthesia Neurodevelopment Assessment.

study that included 997 children who were either never exposed to anesthesia, had a single exposure to general anesthesia, or had multiple exposures to anesthesia before the age of 3 years. Although the study found no differences in IQ scores between groups, a modest reduction in processing speed, fine motor coordination, and a higher incidence of behavioral problems were observed in the group exposed to multiple anesthetics compared with children who were not exposed to any anesthetic before the age of 3 years. The study also observed that children who were exposed to one or multiple anesthetics before the age of 3 years experienced more difficulties with executive function and reading compared to the children who were not exposed to anesthesia. Finally, the Avon Longitudinal Study of Parents and Children study investigated a matched cohort of 13,433 children and compared groups in the same manner as in the MASK study. The authors observed that anesthesia exposure was associated with lower social communication scores for 4 and 9 years old, and lower manual dexterity for 7 years old.

The General Anesthesia compared to Spinal Anesthesia trial, the only randomized controlled trial to date that has aimed to address this topic, randomized 722 infants undergoing inguinal hernia repair to receive either sevoflurane general anesthesia or regional anesthesia with spinal or caudal neuraxial blockade.[32] The authors found that that a brief exposure to sevoflurane anesthesia was not associated with the later development of neurocognitive deficits at 5 years of age.

The lack of clear and reproducible demonstration of anesthesia-induced neurotoxicity in humans has led many experts to question whether the early preclinical trials effectively translate to humans. Several factors might account for the inconsistency demonstrated by human studies. First, the human brain has some inherent differences compared with other species with respect to neurodevelopment and maturation, and is capable of much higher-order cognitive function. Second, the doses and length of exposure anesthetics in the animal models used often far exceeds what is typically used in clinical anesthesia practice, which raises questions to the clinical relevancy of the aforementioned animal models. Third, reproducible markers of neuronal apoptosis that can be used to measure the extent of histological neurodegeneration in humans are lacking, as are reliable neurocognitive or behavioral deficits that accurately reflect neuronal apoptosis. Lastly, many experts have raised significant ethical concerns with respect to conducting double-blinded randomized placebo-controlled trials in young children to assess the later effects of anesthesia-induced neurotoxicity.[33] Taken together, these limitations have greatly hindered the ability to both conduct and interpret clinical studies that examine these effects in humans, and the subject of anesthesia-induced neurotoxicity remains one of the most heavily debated topics in anesthesia practice today.

Exposure to anesthesia and the elderly

In addition to the concerns expressed regarding possible neurotoxic effects of anesthesia exposure on the developing brain, the elderly brain might also be particularly vulnerable to these potential neurotoxic effects. Advanced age is a well-known independent factor that has been associated with poor perioperative outcomes. A recent study in the British Journal of Surgery used a time trend ecological analysis of hospital statistics data for England to demonstrate that the average age of patients undergoing surgery was 47.5 years in 1999 and 54.2 years in 2015, while the average age of the English population aged from 38.3 to 39.7 years over the same time period.[34] The same study predicted that by the year 2030, one-fifth of the population aged 75 years and older will undergo surgery each year. It is therefore very important to elucidate which patients, if any, are particularly vulnerable to neurotoxic effects of anesthesia and to explore possible methods to ameliorate these effects.

Elderly patients are known to experience a higher incidence of perioperative complications compared with younger patients, including postoperative delirium and POCD, which are associated with an increased risk of morbidity and mortality. Age-related anatomical and physiological changes contribute to a potential decrease in cognitive reserve and associated increased susceptibility to neurocognitive decline postoperatively. These age-related changes include a reduction in white matter volume after the age of 45 years,[35] increased blood brain barrier permeability,[36] a gradual reduction in neurogenesis,[37] increase in brain inflammation and production of cytokines,[38,39] cognitive and memory decline,[40,41] a reduction in neurocognitive reserve,[42] and increased risk of cerebrovascular disease.[43] The combined effects of these changes likely contribute to observed increased risk of postoperative delirium and POCD.[44]

Anesthesia exposure in the elderly population has been suggested to possibly accelerate these effects, although this has not been fully elucidated and several recent studies have in fact failed to confirm a definitive link between anesthesia exposure and neurotoxicity. Many experts have suggested that perioperative stress and comorbidities present in the elderly population might play a more important role on the progression of dementia, POCD, and postoperative delirium more so than anesthesia exposure per se.[45–47]

Neuroprotective strategies in the perioperative setting

With the plethora of preclinical evidence that have demonstrated a link between anesthesia exposure and the various mechanisms of anesthesia-induced neurotoxicity explored, several investigators aimed to study whether adjuvant agents can protect against or ameliorate the effects of anesthesia-induced neurotoxicity. Typically, preclinical and clinical studies that have addressed this question have looked at agents that have previously been shown to possess inherent neuroprotective properties or agents that might facilitate a reduction in anesthesia exposure (e.g., dose, concentration, or both).[33] Recent studies have examined the potential neuroprotective effects of centrally acting alpha-2-agonists, rutin, vitamin C, TNF-alpha, and lithium among others.

Dexmedetomidine

The specific alpha-2-agonist studied most frequently is dexmedetomidine, an agent that is approved by the FDA for sedation in the intensive care unit. Dexmedetomidine is also frequently used as an anesthetic adjunct, especially in neurosurgery and enhanced recovery after surgery protocols. Importantly, dexmedetomidine demonstrates less neurotoxicity compared with other anesthetic agents in neonate and pediatric animal models, and recent clinical trials and case reports have even suggested dexmedetomidine might in fact have inherent neuroprotective properties.[48] Although the exact mechanisms are unclear, dexmedetomidine appears to attenuate the neurotoxicity of propofol in rat hippocampal models[49–51] in a manner that is likely dose dependent.[50]

Rutin

Rutin, a citrus flavonoid found in asparagus, buckwheat, and various fruit, has been recently studied in a variety of contexts for its pharmacologic potential uses in preventing neuroinflammation prevention and promoting neural crest cell survival. Rutin has been shown to possess anticonvulsant, antidepressant, and analgesic properties and might potentially slows the effects of AD progression.[52] Rutin was recently shown to attenuate the neurotoxic effects of propofol and sevoflurane in a neonatal rodent model,[53] although this finding needs to be confirmed in future clinical trials.

Vitamin C

Oxidative stress is known to mediate various physiological processes including cell cycle regulation, inflammation, stress responses, and a myriad of second messenger pathways. Collectively, the molecular constituents (e.g., hydrogen peroxide, hydroxyl radical, superoxide radical ion, NO) are known as ROS. Moderate amounts of ROS are necessary for physiological homeostasis, however imbalances of ROS are implicated in the development of various disease processes including cancer, arthritis, diabetes mellitus type I and other autoimmune disorders, neurodegenerative disorders including AD, Parkinson's disease, Huntington's disease, and cardiovascular disease.

The effects of this imbalance of ROS on the brain is quite significant, as the brain demands approximately 20% of oxygen delivery despite comprising only approximately 2% of the total body weight. Moreover, the brain exhibits limited regenerative ability. The development of antioxidant agents in an effort to combat the effects exerted by excess ROS remains an attractive pharmacologic and pharmacogenomic target.

Vitamin C is a particularly attractive agent due to its ubiquitous and inexpensive nature. Vitamin C exerts its antioxidant effects by reducing H_2O_2 to H_2O, regenerating alpha-tocopherol (vitamin E) from its radical. In the brain, vitamin C participates in the biosynthesis of dopamine and norepinephrine, modulation of neurotransmission, and in neurodifferentiation.[54] In one recent study, administration of vitamin C to rats resulted in increased catalase, superoxide dismutase, and NO levels, suggesting that vitamin C might exerts neuroprotective effects by decreasing oxidative stress.[55] In another study, administration of vitamin C to rats resulted in a reduction of anesthesia-induced neuronal apoptosis and subsequent neurocognitive dysfunction.[56]

Modulation of TNF-alpha expression

TNF-alpha is a pleiotropic cytokine which is abnormally present in high concentrations in brain tissue following an acute brain injury. Early studies established that the addition of exogenous TNF-alpha increases the cerebrovascular accident infarct area size, and blocking endogenous TNF-alpha significantly reduces infarct size.[57] More recently, glutamate concentrations were found to be elevated secondary to increased kidney-type glutaminase in rat and human neuronal cell cultures after exogenous TNF-alpha exposure.[58] Moreover, as discussed above, suppression of TNF-alpha expression resulted in decreased propofol-[59] and ketamine-induced[11] neurotoxicity. These findings all support the potential role of targeting glutaminase modulation, and by extension TNF-alpha, as a therapeutic agent for neurodegenerative disease.

In vivo, the aforementioned pathway is thought to occur primarily in microglia during sustained states of neuroinflammation. A recent study identified a novel TNF-alpha-mediated micro(mi)RNA (miR-342) using in vitro miRNA microarrays and quantitative reverse transcription polymerase chain reaction validation of rat microglial cell lines, which increased secretin of TNF-alpha when stimulated.[60] The potential neuroprotective effects of TNF-alpha humans, and its use in the perioperative period to reduce exposure to neurotoxic agents should be further examined by future studies.

Lithium

Although lithium is primarily used to treat refractory depression and bipolar disorder, its potential role as a neuroprotective agent in other disorders has recently been investigated. In laboratory and clinical models, treatment with lithium effected neuronal homeostasis in a variety of mechanisms. In cultured cerebellar granule cells, lithium has been shown to suppress proapoptotic p53 and Bax expression and increase concentrations of cytoprotective protein Bcl-2.[61] Recent evidence has further shown that in addition to reducing oxidative stress,[62] lithium stimulates vascular endothelial growth factor and brain-derive neurotrophic factor, thereby increasing synaptic plasticity and enhancing hippocampal cell survival.[63] Finally, coadministration of lithium with isoflurane for 5 h in neonatal nonhuman primates on postnatal day 6 completely blocked isoflurane-induced-neurotoxicity, establishing a possible clinical role for lithium as a neuroprotective adjuvant agent.[64]

Anesthesia and neurotoxicity: future experimental modalities

Although neurotoxicity has historically been primarily defined by its biological perturbations on a molecular level, the downstream effects manifest not only as widespread and degenerative disease, but also as end-organ damage, perfusion defects, hemorrhage, infarction, calcification, demyelination or atrophy, and brain edema that are mediated by several biological pathways. Recent evidence suggests that neuroimaging modalities and novel fluid-based biomarkers have particular utility in characterizing the extent of neurotoxicity after an acute brain insult.

Neuroimaging

Although the etiology of neurological disease is not always immediately apparent, many effects in the brain can be observed via imaging with magnetic resonance imaging (MRI). MRI is commonly used in neurology and neurosurgery to characterize and differentiate pathophysiology using a combination of techniques such as T2-weighted imaging, fluid-attenuated inversion recovery imaging, and diffusion weighted imaging.[65] MRI can be particularly useful to study longitudinal changes that might reflect neurotoxicity in translational animal models, clinical studies, and postmortem analysis. Other neuroimaging methodologies that can be used in future studies to characterize the extent of neurotoxicity include magnetic resonance spectroscopy and positron emission tomography.

Translational biomarkers of neurotoxicity

Novel biomarkers of neurotoxicity are lacking in current clinical practice but can be extremely useful to diagnose neurotoxicity and monitor the response to treatment. The ability to track disease progression with biomarkers is a cornerstone of current medical practice, and fluid-based biomarkers present in the cerebrospinal fluid represent a particularly attractive biomarker target because of their proximity to neuronal tissues and relative impermeability with other body fluid compartments (Table 4.3).

Conclusions

Numerous preclinical translational studies have used animal models, including nonhuman primates, to demonstrate neurotoxic effects of both volatile and intravenous anesthetics in the brain at both extremes of age, including induction of apoptosis, inhibition of neurogenesis, and proliferation of an inflammatory milieu. These neurotoxic effects in animal models have raised considerable concern in humans, who undergo approximately 40 million anesthetics annually in the United States alone. Although early exposure to anesthetics has been thought to possibly contribute to

Table 4.3 Potential biomarkers of neurotoxicity.

Biomarker	Endpoint
GFAP (glial fibrillary acidic protein)	Biomarker of all types of neural (neuronal and glial) damage
MAP-2 (microtubule-associated protein)	Biomarker of dendritic injury
Microtubule-associated protein tau (total tau, phosphorylated tau, and cleaved tau)	Biomarker of neurodegeneration/ axonal injury
Neurofilament (light chain and phosphorylated heavy chain)	Biomarkers of axonal injury

Notes: Included in this table are various known biomarkers and the potential clinical impact of isolating the individual biomarker in cerebrospinal fluid or serum. These particular biomarkers all have established ELISA data.

later neurocognitive and developmental deficits, and late exposure has been thought to contribute to POCD, clinical trials have demonstrated inconsistent and conflicting results. Future studies should continue to investigate whether anesthesia exposure results in neurotoxicity, as well as determining which phenotypes best reflect this exposure. The novel use of biomarkers and neuroimaging techniques might provide further insight on this important topic.

References

1. Rasmussen LS. Postoperative cognitive dysfunction: incidence and prevention. *Best Pract Res Clin Anaesthesiol*. 2006;20(2):315–330. Available from: https://doi.org/10.1016/j.bpa.2005.10.011.
2. Franks NP, Lieb WR. Molecular and cellular mechanisms of general anaesthesia. *Nature*. 1994;367(6464):607–614. Available from: https://doi.org/10.1038/367607a0.
3. Jevtović-Todorović V, Todorović SM, Mennerick S, et al. Nitrous oxide (laughing gas) is an NMDA antagonist, neuroprotectant and neurotoxin. *Nat Med*. 1998;4(4):460–463. Available from: https://doi.org/10.1038/nm0498-460.
4. Jevtovic-Todorovic V, Hartman RE, Izumi Y, et al. Early exposure to common anesthetic agents causes widespread neurodegeneration in the developing rat brain and persistent learning deficits. *J Neurosci*. 2003;23(3):876–882. Available from: https://doi.org/10.1523/jneurosci.23-03-00876.2003. PMID: 12574416; PMCID: PMC6741934.
5. West MJ. Stereological methods for estimating the total number of neurons and synapses: issues of precision and bias. *Trends Neurosci*. 1999;22(2):51–61. Available from: https://doi.org/10.1016/s0166-2236(98)01362-9.
6. Lu Y, Huang Y, Jiang J, et al. Neuronal apoptosis may not contribute to the long-term cognitive dysfunction induced by a brief exposure to 2% sevoflurane in developing rats. *Biomed Pharmacother*. 2016;78:322–328. Available from: https://doi.org/10.1016/j.biopha.2016.01.034.
7. Xiao H, Liu B, Chen Y, Zhang J. Learning, memory and synaptic plasticity in hippocampus in rats exposed to sevoflurane. *Int J Dev Neurosci*. 2016;48:38–49. Available from: https://doi.org/10.1016/j.ijdevneu.2015.11.001.

8. Tao G, Xue Q, Luo Y, Li G, Xia Y, Yu B. Isoflurane is more deleterious to developing brain than desflurane: the role of the Akt/GSK3β signaling pathway. *Biomed Res Int.* 2016;2016:7919640. Available from: https://doi.org/10.1155/2016/7919640. Epub 2016 Feb 1. PMID: 27057548; PMCID: PMC4753322.
9. Brambrink AM, Evers AS, Avidan MS, Farber NB, Smith DJ, Zhang X, Dissen GA, Creeley CE, Olney JW. Isoflurane-induced neuroapoptosis in the neonatal rhesus macaque brain. *Anesthesiology.* 2010 Apr;112(4):834–841. Available from: https://doi.org/10.1097/ALN.0b013e3181d049cd. PMID: 20234312; PMCID: PMC3962067.
10. Huang H, Liu CM, Sun J, et al. Ketamine affects the neurogenesis of the hippocampal dentate gyrus in 7-day-old rats. *Neurotox Res.* 2016;30(2):185–198. Available from: https://doi.org/10.1007/s12640-016-9615-7.
11. Zheng X, Zhou J, Xia Y. The role of TNF-α in regulating ketamine-induced hippocampal neurotoxicity. *Arch Med Sci.* 2015;11(6):1296–1302. Available from: https://doi.org/10.5114/aoms.2015.56355. Epub 2015 Dec 11. PMID: 26788093; PMCID: PMC4697060.
12. Huang J, Jing S, Chen X, et al. Propofol administration during early postnatal life suppresses hippocampal neurogenesis. *Mol Neurobiol.* 2016;53(2):1031–1044. Available from: https://doi.org/10.1007/s12035-014-9052-7.
13. Milanovic D, Pesic V, Loncarevic-Vasiljkovic N, et al. The Fas ligand/Fas death receptor pathways contribute to propofol-induced apoptosis and neuroinflammation in the brain of neonatal rats. *Neurotox Res.* 2016;30(3):434–452. Available from: https://doi.org/10.1007/s12640-016-9629-1. Epub 2016 May 17. PMID: 27189477.
14. Brambrink A, Orfanakis A. "Therapeutic hypercapnia" after ischemic brain injury: is there a potential for neuroprotection? *Anesthesiology.* 2010;112(2):274–276. Available from: https://doi.org/10.1097/ALN.0b013e3181ca8273.
15. Disma N, Mondardini MC, Terrando N, Absalom AR, Bilotta F. A systematic review of methodology applied during preclinical anesthetic neurotoxicity studies: important issues and lessons relevant to the design of future clinical research. *Paediatr Anaesth.* 2016;26 (1):6–36. Available from: https://doi.org/10.1111/pan.12786.
16. Bosnjak ZJ, Logan S, Liu Y, Bai X. Recent insights into molecular mechanisms of propofol-induced developmental neurotoxicity: implications for the protective strategies. *Anesth Analg.* 2016;123(5):1286–1296. Available from: https://doi.org/10.1213/ane.0000000000001544.
17. Perouansky M, Hemmings Jr. HC. Neurotoxicity of general anesthetics: cause for concern? *Anesthesiology.* 2009;111(6):1365–1371. Available from: https://doi.org/10.1097/ALN.0b013e3181bf1d61.
18. Jevtovic-Todorovic V. Functional implications of an early exposure to general anesthesia: are we changing the behavior of our children? *Mol Neurobiol.* 2013;48(2):288–293. Available from: https://doi.org/10.1007/s12035-013-8488-5.
19. Vutskits L, Xie Z. Lasting impact of general anaesthesia on the brain: mechanisms and relevance. *Nat Rev Neurosci.* 2016;17(11):705–717. Available from: https://doi.org/10.1038/nrn.2016.128.
20. Sanchez V, Feinstein SD, Lunardi N, et al. General anesthesia causes long-term impairment of mitochondrial morphogenesis and synaptic transmission in developing rat brain. *Anesthesiology.* 2011;115(5):992–1002. Available from: https://doi.org/10.1097/ALN.0b013e3182303a63.
21. Brambrink AM, Orfanakis A, Kirsch JR. Anesthetic neurotoxicity. *Anesthesiol Clin.* 2012;30(2):207–228. Available from: https://doi.org/10.1016/j.anclin.2012.06.002.

22. Culley DJ, Baxter MG, Crosby CA, Yukhananov R, Crosby G. Impaired acquisition of spatial memory 2 weeks after isoflurane and isoflurane-nitrous oxide anesthesia in aged rats. *Anesth Analg*. 2004;99(5):1393–1397. Available from: https://doi.org/10.1213/01.Ane.0000135408.14319.Cc.
23. Xie Z, Culley DJ, Dong Y, et al. The common inhalation anesthetic isoflurane induces caspase activation and increases amyloid beta-protein level in vivo. *Ann Neurol*. 2008;64(6):618–627. Available from: https://doi.org/10.1002/ana.21548.
24. Roberson ED, Scearce-Levie K, Palop JJ, et al. Reducing endogenous tau ameliorates amyloid beta-induced deficits in an Alzheimer's disease mouse model. *Science*. 2007;316(5825):750–754. Available from: https://doi.org/10.1126/science.1141736.
25. Graham MR, Brownell M, Chateau DG, Dragan RD, Burchill C, Fransoo RR. Neurodevelopmental assessment in Kindergarten in children exposed to general anesthesia before the age of 4 years: a retrospective matched cohort study. *Anesthesiology*. 2016;125(4):667–677. Available from: https://doi.org/10.1097/aln.0000000000001245.
26. O'Leary JD, Janus M, Duku E, et al. A population-based study evaluating the association between surgery in early life and child development at primary school entry. *Anesthesiology*. 2016;125(2):272–279. Available from: https://doi.org/10.1097/aln.0000000000001200. PMID: 27433745.
27. O'Leary JD, Orser BA. Neurodevelopment after general anaesthesia in infants. *Lancet*. 2019;393(10172):614–615. Available from: https://doi.org/10.1016/s0140-6736(18)32985-4.
28. Sun LS, Li G, Miller TL, Salorio C, Byrne MW, Bellinger DC, et al. Association between a single general anesthesia exposure before age 36 months and neurocognitive outcomes in later childhood. *JAMA*. 2016;315(21):2312–2320. Available from: https://doi.org/10.1001/jama.2016.6967. PMID: 27272582; PMCID: PMC5316422.
29. Warner DO, Zaccariello MJ, Katusic SK, Schroeder DR, Hanson AC, Schulte PJ, et al. Neuropsychological and behavioral outcomes after exposure of young children to procedures requiring general anesthesia: the mayo anesthesia safety in kids(MASK) study. *Anesthesiology*. 2018;129(1):89–105. Available from: https://doi.org/10.1097/ALN.0000000000002232. PMID: 29672337; PMCID: PMC6008202.
30. Walkden GJ, Gill H, Davies NM, Peters AE, Wright I, Pickering AE. Early childhood general anesthesia and neurodevelopmental outcomes in the avon longitudinal study of parents and children birth cohort. *Anesthesiology*. 2020;133(5):1007–1020. Available from: https://doi.org/10.1097/ALN.0000000000003522. PMID: 32898216.
31. McCann ME, de Graaff JC, Dorris L, Disma N, Withington D, Bell G, et al. G.A.S. Consortium. Neurodevelopmental outcome at 5 years of age after general anaesthesia or awake-regional anaesthesia in infancy (GAS): an international, multicentre, randomised, controlled equivalence trial. *Lancet*. 2019;393(10172):664–677. Available from: https://doi.org/10.1016/S0140-6736(18)32485-1. Epub 2019 Feb 14. Erratum in: Lancet. 2019 Aug 24;394(10199):638. PMID: 30782342; PMCID: PMC6500739.
32. Davidson AJ, Disma N, de Graaff JC, et al. Neurodevelopmental outcome at 2 years of age after general anaesthesia and awake-regional anaesthesia in infancy (GAS): an international multicentre, randomised controlled trial. *Lancet*. 2016;387(10015):239–250. Available from: https://doi.org/10.1016/s0140-6736(15)00608-x.
33. Bilotta F, Evered LA, Gruenbaum SE. Neurotoxicity of anesthetic drugs: an update. *Curr Opin Anaesthesiol*. 2017;30(4):452–457. Available from: https://doi.org/10.1097/aco.0000000000000482.
34. Fowler AJ, Abbott TEF, Prowle J, Pearse RM. Age of patients undergoing surgery. *Br J Surg*. 2019;106(8):1012–1018. Available from: https://doi.org/10.1002/bjs.11148.

35. Aine CJ, Sanfratello L, Adair JC, Knoefel JE, Caprihan A, Stephen JM. Development and decline of memory functions in normal, pathological and healthy successful aging. *Brain Topogr.* 2011;24(3):323–339. Available from: https://doi.org/10.1007/s10548-011-0178-x.
36. Zeevi N, Pachter J, McCullough LD, Wolfson L, Kuchel GA. The blood-brain barrier: geriatric relevance of a critical brain-body interface. *J Am Geriatr Soc.* 2010;58 (9):1749–1757. Available from: https://doi.org/10.1111/j.1532-5415.2010.03011.x.
37. Couillard-Despres S, Iglseder B, Aigner L. Neurogenesis, cellular plasticity and cognition: the impact of stem cells in the adult and aging brain—a mini-review. *Gerontology.* 2011;57(6):559–564. Available from: https://doi.org/10.1159/000323481.
38. Corona AW, Fenn AM, Godbout JP. Cognitive and behavioral consequences of impaired immunoregulation in aging. *J Neuroimmune Pharmacol.* 2012;7(1):7–23. Available from: https://doi.org/10.1007/s11481-011-9313-4.
39. Ownby RL. Neuroinflammation and cognitive aging. *Curr Psychiatry Rep.* 2010;12 (1):39–45. Available from: https://doi.org/10.1007/s11920-009-0082-1.
40. Salthouse TA. Neuroanatomical substrates of age-related cognitive decline. *Psychol Bull.* 2011;137(5):753–784. Available from: https://doi.org/10.1037/a0023262.
41. Small SA. Age-related memory decline: current concepts and future directions. *Arch Neurol.* 2001;58(3):360–364. Available from: https://doi.org/10.1001/archneur.58.3.360.
42. Stern Y. Cognitive reserve. *Neuropsychologia.* 2009;47(10):2015–2028. Available from: https://doi.org/10.1016/j.neuropsychologia.2009.03.004.
43. Grinberg LT, Thal DR. Vascular pathology in the aged human brain. *Acta Neuropathol.* 2010;119(3):277–290. Available from: https://doi.org/10.1007/s00401-010-0652-7.
44. Strøm C, Rasmussen LS, Sieber FE. Should general anaesthesia be avoided in the elderly? *Anaesthesia.* 2014;69(Suppl 1):35–44. Available from: https://doi.org/10.1111/anae.12493.
45. Jevtovic-Todorovic V. General anesthetics and neurotoxicity: how much do we know? *Anesthesiol Clin.* 2016;34(3):439–451. Available from: https://doi.org/10.1016/j.anclin.2016.04.001.
46. Armstrong R, Xu F, Arora A, Rasic N, Syed NI. General anesthetics and cytotoxicity: possible implications for brain health. *Drug Chem Toxicol.* 2017;40(2):241–249. Available from: https://doi.org/10.1080/01480545.2016.1188306.
47. Bittner EA, Yue Y, Xie Z. Brief review: anesthetic neurotoxicity in the elderly, cognitive dysfunction and Alzheimer's disease. *Can J Anaesth.* 2011;58(2):216–223. Available from: https://doi.org/10.1007/s12630-010-9418-x.
48. Alam A, Suen KC, Hana Z, Sanders RD, Maze M, Ma D. Neuroprotection and neurotoxicity in the developing brain: an update on the effects of dexmedetomidine and xenon. *Neurotoxicol Teratol.* 2017;60:102–116. Available from: https://doi.org/10.1016/j.ntt.2017.01.001.
49. Li J, Xiong M, Nadavaluru PR, et al. Dexmedetomidine attenuates neurotoxicity induced by prenatal propofol exposure. *J Neurosurg Anesthesiol.* 2016;28(1):51–64. Available from: https://doi.org/10.1097/ana.0000000000000181.
50. Liu H, Wang T, Dai W, Jiang Z, Li YH, Liu XS. Subhypnotic doses of propofol impair spatial memory retrieval in rats. *Neural Regen Res.* 2016;11(12):1956–1961. Available from: https://doi.org/10.4103/1673-5374.197137.
51. Wei Y, Hu J, Liang Y, et al. Dexmedetomidine pretreatment attenuates propofol-induced neurotoxicity in neuronal cultures from the rat hippocampus. *Mol Med Rep.* 2016;14 (4):3413–3420. Available from: https://doi.org/10.3892/mmr.2016.5628.

52. Ganeshpurkar A, Saluja AK. The pharmacological potential of rutin. *Saudi Pharm J*. 2017;25(2):149–164. Available from: https://doi.org/10.1016/j.jsps.2016.04.025.
53. Man YG, Zhou RG, Zhao B. Efficacy of rutin in inhibiting neuronal apoptosis and cognitive disturbances in sevoflurane or propofol exposed neonatal mice. *Int J Clin Exp Med*. 2015;8(8):14397–14409.
54. Teleanu RI, Chircov C, Grumezescu AM, Volceanov A, Teleanu DM. Antioxidant therapies for neuroprotection-a review. *J Clin Med*. 2019;8(10):1659. Available from: https://doi.org/10.3390/jcm8101659.
55. Sangeetha S, Murali M. Non-covalent DNA binding, protein interaction, DNA cleavage and cytotoxicity of [Cu(quamol)Cl]·H(2)O. *Int J Biol Macromol*. 2018;107(Pt B):2501–2511. Available from: https://doi.org/10.1016/j.ijbiomac.2017.10.131.
56. Xu KX, Tao J, Zhang N, Wang JZ. Neuroprotective properties of vitamin C on equipotent anesthetic concentrations of desflurane, isoflurane, or sevoflurane in high fat diet fed neonatal mice. *Int J Clin Exp Med*. 2015;8(7):10444–10458.
57. Barone FC, Arvin B, White RF, et al. Tumor necrosis factor-alpha. A mediator of focal ischemic brain injury. *Stroke*. 1997;28(6):1233–1244. Available from: https://doi.org/10.1161/01.str.28.6.1233.
58. Ye L, Huang Y, Zhao L, et al. IL-1β and TNF-α induce neurotoxicity through glutamate production: a potential role for neuronal glutaminase. *J Neurochem*. 2013;125(6):897–908. Available from: https://doi.org/10.1111/jnc.12263.
59. Chen B, Deng X, Wang B, Liu H. Etanercept, an inhibitor of TNF-a, prevents propofol-induced neurotoxicity in the developing brain. *Int J Dev Neurosci*. 2016;55:91–100. Available from: https://doi.org/10.1016/j.ijdevneu.2016.10.002.
60. Brás JP, Bravo J, Freitas J, et al. TNF-alpha-induced microglia activation requires miR-342: impact on NF-kB signaling and neurotoxicity. *Cell Death Dis*. 2020;11(6):415. Available from: https://doi.org/10.1038/s41419-020-2626-6.
61. Chen RW, Chuang DM. Long term lithium treatment suppresses p53 and Bax expression but increases Bcl-2 expression. A prominent role in neuroprotection against excitotoxicity. *J Biol Chem*. 1999;274(10):6039–6042. Available from: https://doi.org/10.1074/jbc.274.10.6039.
62. Shalbuyeva N, Brustovetsky T, Brustovetsky N. Lithium desensitizes brain mitochondria to calcium, antagonizes permeability transition, and diminishes cytochrome c release*. *J Biol Chem*. 2007;282(25):18057–18068. Available from: https://doi.org/10.1074/jbc.M702134200.
63. Diniz BS, Teixeira AL. Brain-derived neurotrophic factor and Alzheimer's disease: physiopathology and beyond. *NeuroMolecular Med*. 2011;13(4):217–222. Available from: https://doi.org/10.1007/s12017-011-8154-x.
64. Noguchi KK, Johnson SA, Kristich LE, et al. Lithium protects against anaesthesia neurotoxicity in the infant primate brain. *Sci Rep*. 2016;6:22427. Available from: https://doi.org/10.1038/srep22427.
65. Roberts RA, Aschner M, Calligaro D, et al. Translational biomarkers of neurotoxicity: a Health and Environmental Sciences Institute perspective on the way forward. *Toxicol Sci*. 2015;148(2):332–340. Available from: https://doi.org/10.1093/toxsci/kfv188.

Neuromonitoring and translational research

5

Ankur Luthra, Summit Bloria, Rajeev Chauhan and Sameer Sethi
Department of Anaesthesia and Intensive Care, PGIMER, Chandigarh, India

Introduction

Neuromonitoring refers to the use of various modalities to monitor the integrity of the nervous system. Unlike the other body systems that have been easy to monitor, evaluation of structural integrity and function of the nervous system has been difficult. However, over last two decades, there has been a significant upward trend in development of newer devices for neuromonitoring. The modalities of neuromonitoring are used in operation theaters as well as in the intensive care unit (ICU). In this chapter, we touch upon all the modalities of neuromonitoring as well as the evidence supporting their effectiveness. The following modalities have been used for the purpose of neuromonitoring:

1. Clinical monitoring
2. Imaging studies
3. Intracranial pressure (ICP) and derived indices
4. Optic nerve sheath diameter (ONSD)
5. Transcranial Doppler (TCD)
6. Pupillometry
7. Thermal diffusion flowmetry
8. Jugular venous oxygen saturation
9. Brain tissue oxygen partial pressure
10. Near infrared spectroscopy (NIRS)
11. Electroencephalography
12. Evoked potentials (EPs).

Clinical monitoring

Before the advent of fancy devices, clinical evaluation of the patient was the only means of determination of neurological status of patient. Various scales have been suggested and used for evaluation of neurological status of patients.

Glasgow coma score

Jennet and Teasdale developed the Glasgow coma scale (GCS) for standardization of the patients with traumatic brain injury (TBI). GCS has been used to classify the

Perioperative Neuroscience. DOI: https://doi.org/10.1016/B978-0-323-91003-3.00007-6

TBI as per severity, identify changes in status of patients, and guide in their prognostication (Table 5.1).[1,2] It has also been found effective in prognostication of patients after trauma.[3] Using GCS scores to classify injuries as mild TBI (GCS scores 13–15), moderate TBI (GCS scores 9–12), or severe TBI (GCS scores ≤ 8) is an accepted norm. However, the scale has its limitations—the scale is less useful in intubated patients as verbal responses cannot be assessed in them, GCS does not allow assessment of brainstem function and a GCS score of 3 may cover a spectrum of brain injury severity.

Combined GCS–pupils score

The combined GCS–pupils score (GCS-P) can vary from 1 to 15. GCS-P is calculated by subtracting a pupil reactivity score (PRS) from the calculated GCS. PRS depends upon the number of nonreactive pupils. PRS is 2 if both the pupils are unreactive to light. If only one pupil is unreactive to light the score is 1. The score is 0 if neither pupil is unreactive to light. GCS-P thus can vary from 1 to 15.

Full Outline of Unresponsiveness score

It assesses four components of neurologic function—eye, motor, brainstem, and respiratory functions. It was developed to overcome some of the limitations of GCS (Table 5.1).[4] It has been validated as a good predictor of outcome even in medical ICU patients, pediatric patients, and stroke patients.[5–7] However, Eken et al. did not find Full Outline of Unresponsiveness (FOUR) score to be better than GCS in an emergency setting population.[8] Similarly, Suresh et al. did not find better interobserver variability with the FOUR score as compared to GCS.[9] Also, it has not been proven to be better than GCS in pediatric population.[10]

For aneurysmal subarachnoid hemorrhage patients

Modified Hess–Hunt scale (Table 5.2): Hess and Hunt modified the Botterell's classification of aneurysmal subarachnoid hemorrhage (SAH) and suggested that while prompt surgical intervention was beneficial in patients admitted with good condition, conservative therapy until improvement in patient's condition is better for the more seriously ill patients.[11]

World Federation of Neurological Surgeon scale (Table 5.2): Described in 1988, the World Federation of Neurological Surgeons scale is based on GCS.[12]

Fisher scale (Table 5.2): The grading predicts the incidence of vasospasm in aneurysmal SAH patients.[13]

Modified Fisher scale (Table 5.2): It was suggested in 2006, the modified Fisher scale is said to predict symptomatic vasospasm after SAH more accurately than original Fisher scale.[14]

Table 5.1 Glasgow Coma Scale and Full Outline of Unresponsiveness score.

Glasgow Coma Scale
Eye response (E) There are four grades: E1—No opening of the eye E2—Eye opening in response to pain stimulus E3—Eye opening to speech E4—Eyes opening spontaneously
Verbal response (V) There are five grades: V1—No verbal response V2—Incomprehensible sounds V3—Inappropriate words V4—Confused V5—Oriented
Motor response (M) There are six grades: M1—No motor response M2—Abnormal extension in response to pain M3—Abnormal flexion in response to pain M4—Normal flexion M5—Localizes to pain M6—Obeys commands
Full Outline of Unresponsiveness score
Eye response Eyelids open or opened, tracking, or blinking to command—4 Eyelids open but not tracking—3 Eyelids closed but open to loud voice—2 Eyelids closed but open to pain—1 Eyelids remain closed with pain—0
Motor response Thumbs-up, fist, or peace sign—4 Localizing to pain—3 Flexion response to pain—2 Extension response to pain—1 No response to pain or generalized myoclonus status—0
Brainstem reflexes Pupil and corneal reflexes present—4 One pupil wide and fixed—3 Pupil or corneal reflexes absent—2 Pupil and corneal reflexes absent—1 Absent pupil, corneal, and cough reflex—0

(*Continued*)

Table 5.1 (Continued)

Respiration
Not intubated, regular breathing pattern—4
Not intubated, Cheyne–Stokes breathing pattern—3
Not intubated, irregular breathing—2
Breathes above ventilator rate—1
Breathes at ventilator rate or apnea—0

Table 5.2 Modified Hess–Hunt scale, World Federation of Neurological Surgeons scale, Fisher and modified Fisher scales.

Modified Hess–Hunt scale
There are six grades: 0—Unruptured aneurysm I—Asymptomatic or minimal headache and slight nuchal rigidity II—Moderate to severe headache, nuchal rigidity, but no neurologic deficit other than cranial nerve palsy III—Drowsiness, confusion, or mild focal deficit IV—Stupor, mild or severe hemiparesis, possible early decerebrate rigidity, vegetative disturbance V—Deep coma, decerebrate rigidity, moribund appearance
World Federation of Neurological Surgeon scale
There are five grades: Grade 1: GCS 15, no motor deficitGrade 2: GCS 13–14, without deficitGrade 3: GCS 13–14, with focal neurological deficitGrade 4: GCS 7–12, with or without deficitGrade 5: GCS < 7, with or without deficit
Fisher scale
There are four grades: 1. No blood detected 2. Diffuse thin layer of subarachnoid blood (vertical layers <1 mm thick) 3. Localized clot or thick layer of subarachnoid blood (vertical layers ≥ 1 mm thick) 4. Intracerebral or intraventricular blood with diffuse or no subarachnoid blood
Modified Fisher scale
Grade 0 No subarachnoid hemorrhage (SAH)/No intraventricular hemorrhage (IVH) Grade 1 Focal or diffuse, thin SAHNo IVH Grade 2 Thin focal or diffuse SAH/IVH present Grade 3 Thick focal or diffuse SAHNo IVH Grade 4 Thick focal or diffuse SAH/IVH present

For ischemic stroke patients

National Institutes of Health Stroke Scale (NIHSS) (Table 5.3) is used to classify patients of ischemic stroke into mild, moderate, and severe stroke. It "provides a

Table 5.3 National Institutes of Health Stroke scale.

1a—Level of consciousness	0 = Alert; keenly responsive 1 = Not alert, but arousable by minor stimulation 2 = Not alert; requires repeated stimulation 3 = Unresponsive or responds only with reflex
1b—Level of consciousness questions: What is your age? What is the month?	0 = Answers two questions correctly 1 = Answers one question correctly 2 = Answers neither question correctly
1c—Level of consciousness commands: Open and close your eyes Grip and release your hand	0 = Performs both tasks correctly 1 = Performs one task correctly 2 = Performs neither task correctly
2—Best gaze	0 = Normal 1 = Partial gaze palsy 2 = Forced deviation
3—Visual	0 = No visual lost 1 = Partial hemianopia 2 = Complete hemianopia 3 = Bilateral hemianopia
4—Facial palsy	0 = Normal symmetric movements 1 = Minor paralysis 2 = Partial paralysis 3 = Complete paralysis of one or both sides
5—Motor arm Left arm Right arm	0 = No drift 1 = Drift 2 = Some effort against gravity 3 = No effort against gravity 4 = No movement
6—Motor leg Left leg Right leg	0 = No drift 1 = Drift 2 = Some effort against gravity 3 = No effort against gravity 4 = No movement
7—Limb ataxia	0 = Absent 1 = Present in one limb 2 = Present in two limbs

(Continued)

Table 5.3 (Continued)

8—Sensory	0 = Normal; no sensory loss 1 = Mild-to-moderate sensory loss 2 = Severe-to-total sensory loss
9—Best language	0 = No aphasia; normal 1 = Mild-to-moderate aphasia 2 = Severe aphasia 3 = Mute: global aphasia
10—Dysarthria	0 = Normal 1 = Mild-to-moderate dysarthria 2 = Severe dysarthria
11—Extinction and inattention	0 = No abnormality 1 = Visual, tactile, auditory, spatial, or personal inattention 2 = Profound hemi-inattention or extinction
Score = 0–42	

quantitative measure of stroke-related neurologic deficit." The scale consists of 11 different elements, each of which evaluates specific ability. The score for each ability is a number between 0 and 4, with a score of 0 being normal functioning and a score of 4 being completely impaired. The patient's NIHSS score is calculated by adding the number for each element of the scale; 42 is the highest score possible. In the NIHSS, the higher the score, the more impaired a stroke patient is considered.

A NIHSS score over 16 suggests a poorer outcome, while recovery is good in patients with a score of 6 or lesser. It has been suggested that each 1-point increase on the scale decreases the possibility of a good outcome for the patient by 17%.

Imaging studies

Noncontrast computerized tomography

It is the initial investigation in acute TBI and stroke because it accurately predicts sensitivity to hemorrhage with good speed and accessibility. In case of stroke, noncontrast computerized tomography (CT) scan is utilized to differentiate ischemic from hemorrhagic stroke.

In TBI patients, *Marshall classification* and *Rotterdam score* (Table 5.4) relate imaging findings on noncontrast CT scans to predict mortality and patient outcome. Assessment of changes in Rotterdam CT scores over time have been used as a prognostic indicator in TBI and can help guide decision of undergoing decompression craniectomy.[15] In Rotterdam scoring system, 1 point is added as a sum score to make the Rotterdam grade numerically total 6 points, consistent with the motor

Table 5.4 Marshall classification and Rotterdam score.

Marshall class	Characteristics
I	No intracranial pathology seen with computerized tomography
II	Cisterns present with midline shift 0–5 mm and/or lesions/densities present; no high or mixed density lesions >25 cm^3; may include bone fragments and foreign bodies
III	Cisterns compressed or absent with midline shift of 0–5 mm; no high or mixed density lesions >25 cm^3
IV	Midline shift >5 mm; no high or mixed density lesions >25 cm^3
V	Any lesion surgically evacuated
VI	High or mixed density lesion >25 cm^3; not surgically evacuated
Rotterdam score	Predictor value
	Basal cisterns
0	Normal
1	Compressed
2	Absent
	Midline shift
0	No or <5 mmShift >5 mm
1	
	Epidural mass lesion
0	Present
1	Absent
	Intraventricular blood or subarachnoid hemorrhage
0	Absent
1	Present
+1	Sum score
Rotterdam score	Mortality (%)
1	0
2	6.8
3	16
4	26
5	53
6	61

score of GCS and Marshall classification. The Rotterdam score has also been suggested as a predictor of outcome in pediatric patients with TBI.[16] Recently, a new scale called TBI score was proposed for patients of TBI having intracranial hemorrhage.[17] TBI score was obtained with the following calculation: (verbal component of GCS + motor component of GCS) − (epidural component of Rotterdam CT score + intraventricular or SAH components of Rotterdam CT score).

Individual predictors of bad outcome on conventional CT in TBI patients include midline shift, presence of SAH, compressed or absent basilar cisterns, and presence of intracranial hemorrhagic lesions (e.g., subdural or epidural hematomas, shear injury, and contusions).

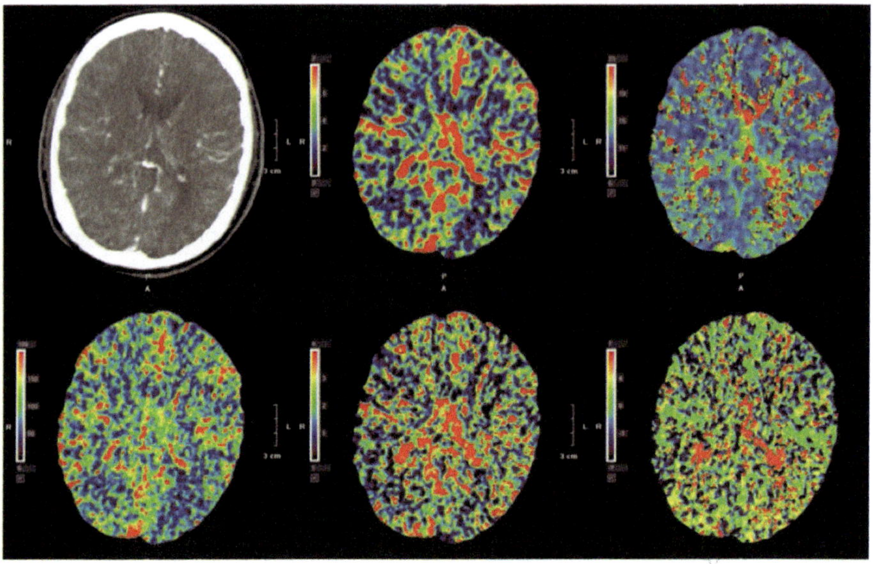

Figure 5.1 Computerized tomography perfusion (normal scan).

Perfusion computerized tomography

It makes use of dynamic scanning after injection of intravenous contrast and illustrates physiologic parameters like cerebral blood volume, cerebral blood flow (CBF), and mean transit time (time for blood to perfuse a region of tissue) (Fig. 5.1). Perfusion CT detects normal brain perfusion or hyperemia or oligemia and reflects the status of cerebrovascular autoregulation.

Magnetic resonance imaging

In TBI patients, magnetic resonance imaging (MRI) is warranted if the CT head does not demonstrate a finding which is in accordance to the clinical status of the patient. Except for SAH and skull fracture, MRI is said to be superior to CT head as an imaging modality. Diffusion tensor imaging is sensitive in detecting diffuse axonal injury. Susceptibility-weighted imaging is exquisitely sensitive in detecting micro-hemorrhages, small veins, and changes in iron content. In magnetoencephalography, "slow waves" are seen in cortical gray matter that is functionally and structurally connected to white matter fiber tracts with axonal shear injury. Functional MRI detects a blood oxygen level-dependent signal and provides information about cerebral function. Disadvantages of imaging studies include the fact that they only provide "snapshot" images of the brain and therefore cannot be used to track the course of brain injury over time or guide neuroprotective treatment strategies in real time. Also, they require shifting of patient to specialized imaging facilities and have a prolonged scanning time, both of which can prove be detrimental to the patient's condition.

Intracranial pressure and derived indices

ICP monitoring allows calculation of absolute ICP measurement, calculation of cerebral perfusion pressure (CPP), and waveform analysis by assessment of cerebrovascular reactivity along with autoregulatory status of the patient. Two methods of monitoring ICP are commonly used in clinical practice. A metaanalysis confirmed that the average error between ICP measures is clinically negligible.[18]

External ventricular catheters

They are considered to be the gold standard for measurement of ICP. Insertion of ventricular catheter employs using surface landmarks to create a burr hole which is placed at Kocher's point. Once cerebrospinal fluid's (CSF) flow is visualized after removal of catheter stylet, it is transduced to obtain an opening ICP. In addition to measuring ICP (which has prognostic value), ventricular catheters can also be used to drain CSF, to withdraw CSF for the purposes of cultures and/or obtaining malignant cells, or to instil medications. The complications associated with use of ventricular catheters include meningitis, catheter obstruction, and bleeding. Intermittent drainage of CSF and rapid weaning of ventricular catheter have been shown to be associated with lesser ventriculoperitoneal shunt placements, fewer complications, and shorter hospital stay in aneurysmal SAH patients.[19] In TBI patients, the use of ventricular catheters has been found to be associated with unfavorable functional and neuropsychological outcomes and higher mortality than intraparenchymal monitors.[20]

The ventricular catheters measure global ICP. Its advantages include low-cost, high accuracy, and reliability and they also allow therapeutic CSF drainage to manage intracranial hypertension, thus can be therapeutic apart from being diagnostic.

However, the disadvantages include limited ability to obtain continuous ICP and difficulty in correct placement in patients with severe cerebral edema along with higher hemorrhage and infection rates and frequent dysfunction/occlusion caused by blood products.

Parenchymal microtransducer devices

Camino transducers

The device has a fibreoptic transducer with a diaphragm and a microprocessor driven amplifier. Changes in light signal intensity are used to calculate ICP.

Codman microsensor

The Codman microsensor (CMS) ICP transducer is a catheter with a microminiature strain gage pressure sensor mounted at one end and an electrical connector at the other end. The risk for hemorrhagic and infectious complications when using the CMS for ICP monitoring is said to be low.[21]

Spiegelberg ICP monitor

It uses pneumatic strain gauze technology. A balloon tipped catheter system is used to measure ICP. When compared with ventricular pressure, the monitor has been proven to provide an accurate measure of ICP.[22]

The biggest advantages of parenchymal devices are that they allow continuous monitoring of ICP and are relatively easy to place specially in the settings of gross cerebral edema. They also have lower complication rates for infection and hemorrhage. However, few disadvantages include their ability to potentially drift over time and the inability of the catheter to be rezeroed once placed. Also, their high cost of monitoring precludes its use at most of the centers. Sometimes, the measured pressure may not represent the true global CSF pressure, as pressure gradients exist between the different brain compartments.

Normal ICPs varies from 7 to 15 mmHg in a supine adult. Raised ICP is a long-established and important cause of TBI-related secondary brain injury and has been associated with higher mortality and poor long-term functional outcomes in many studies. The Brain Trauma Foundation recommends treatment of ICP of greater than or equal to 22 mmHg based on evidence that ICP-guided management of severe TBI may reduce in-hospital and 2-week postinjury mortality.

Chestnut et al. found comparable 3- and 6-month outcomes in patients when treatment was guided by ICP monitoring compared to treatment guided by imaging and clinical examination in the absence of ICP monitoring.[23] However, it is suggested that treatment interventions can be better optimized by individualized interpretation of ICP values in association with other neuromonitoring variables. Also, ICP data provide early warning of deterioration and more accurate prognostication of TBI patients. In a retrospective analysis of 817 TBI patients, an automated computer algorithm was able to predict ICP crises with 30 min advance warning from previous ICP measurements and time since last episode of elevated ICP.[24]

Optic nerve sheath diameter

Chronic ICP elevation affects the optic nerve morphology. Measurement of optic nerve diameter using ultrasound at 3 mm from its origin has been suggested as a means to rule out intracranial hypertension (Fig. 5.2).

ONSD measurements below 5 mm signify normal ICP. An ONSD of greater than 6 mm represents an elevated ICP (ICP $>$ 20 mmHg). The use of ONSD has been shown to be useful in identifying patients with elevated ICP coming to emergency department.[25,26] Newman et al. demonstrated that ONSD was useful in the assessment and monitoring of children suffering with hydrocephalus and suspected of having raised ICP.[27] A reliable correlation between ultrasound and brain CT scan for the diagnosis of ICP rising through measuring ONSD in patients with head trauma has been demonstrated.[28] ONSD measurements are useful for ruling out raised ICP, with sensitivity and specificity approaching 90% and 85%.[29,30] Significant interrater variability is a problem with ONSD, as the procedure involves measuring a structure of few millimeters size, that too behind the eyeball. To

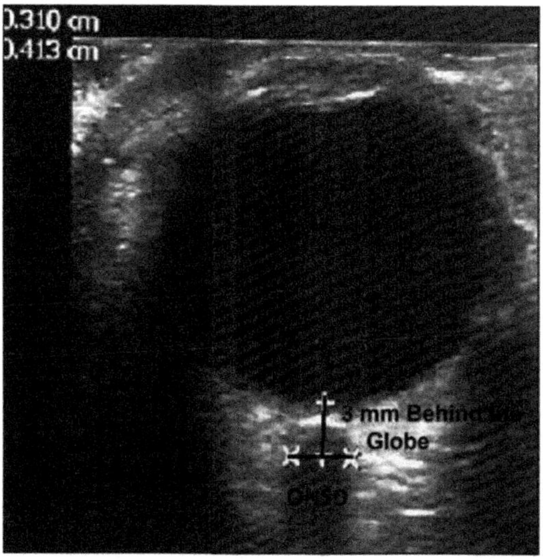

Figure 5.2 Ultrasound image of eye demonstrating eyeball and optic nerve originating from it.

overcome this problem, taking multiple measurements of ONSD over a period of time is recommended. The values thus obtained should be averaged to find out a mean value which is considered to be more representative of a true diameter, thus eliminating error.

Transcranial Doppler

TCD works by measuring the blood flow velocity through intracranial vessels. Visualization of waveforms and flow velocities obtained by TCD is used to qualitatively estimate ICP and CPP (low, normal, and high). The pulsatility index (PI) is an index determined using blood flow velocities and has been shown to strongly correlate with ICP (Fig. 5.3). All methods of TCD-derived PI are based on the fact that ICP and PI correlate positively with increases of ICP. PI values between 1.26 and 1.35 are highly sensitive and specific for ICP >20 mmHg.

TCD has been used to determine the onset of cerebral vasospasm in aneurysmal SAH patients, to check the status of cerebral autoregulation in neurologically ill patients, to identify cerebral arterial occlusion in stroke patients, in eclampsia patients, and in orthoptic liver transplantation.[31–34]

Pupillometry

Elevated ICP/herniation/edema can affect neuroanatomical pathways and lead to an abnormal pupillary response. Infrared pupillometry quantitatively analyzes changes in ICP by examining subtle changes in pupil size in response to different light

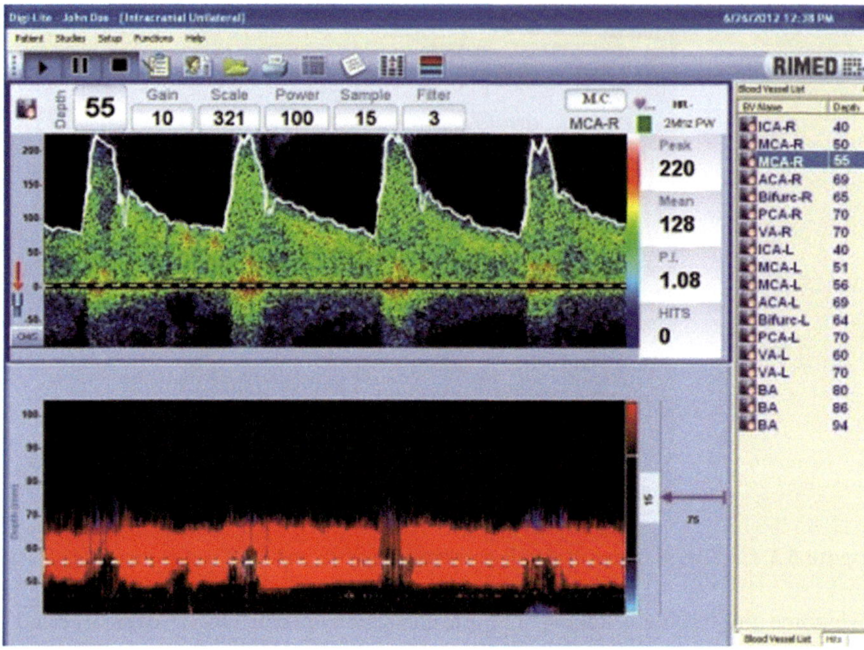

Figure 5.3 Transcranial Doppler view showing pulsatility index.

stimuli. In states of normal ICP, the pupillary response to light should be a 34%–36% decrease in size. High ICP (> 20 mmHg) has been correlated with a pupil response that is reduced by 20% of normal constriction. Neurologic pupillary index (NPi) includes multiple variables pupil size, constriction, dilation velocities, and latency of the pupils. These measurements are then normalized into a standardized score which is compared to standardized scores for healthy controls. A NPi score of 3 or higher indicates a normal pupillary response, while a score of <3 suggests an abnormal pupillary reflex. NPi is helpful in identifying patients who require emergent treatment for potentially fatal ICP levels. Pupillometry is a useful tool for early detection and intervention of many pathologic increases in ICP, as abnormal pupillary activity precedes peak ICP by a difference of hours. NPi has been shown to be predictive of prognosis after cardiac arrest and after venoarterial extracorporeal membrane oxygenation.[35,36] A decrease in NPi has been shown to be correlated with elevated ICP in both adult and pediatric TBI patients.[37,38] In patients with ischemic and hemorrhagic strokes, a significant correlation between septum pellucidum horizontal shift (SPS) and the NPi, constriction velocity of pupil (CV) and pupillary asymmetry has been demonstrated.[39]

Thermal diffusion flowmetry

Thermal diffusion flowmetry is a monitor of regional CBF. It is invasive, continuous, and quantitative. The probe for thermal diffusion flowmetry consists of two gold sensors, one probe transmits heat and other probe collects the emitted heat. The regional CBF determines the thermal difference between the two probes. The temperature difference between the two is a reflection of heat transfer that is converted into an absolute measurement of blood flow in mL/100 g/min. The probe for thermal diffusion flowmetry is inserted into the brain via a burr hole. A region at risk for cerebral ischemia is selected for probe insertion and measurements obtained rely on thermistor temperature dissipation within the region. The probe is usually placed in subcortical white matter at a depth of 20–25 mm. Thermal diffusion flowmetry has been utilized to detect changes in CBF in real time, detect vasospasm in comatose patients, and assess status of cerebral autoregulation. However, there is very limited clinical data using this technology. The device also needs recalibration at regular intervals. Also, there have been concerns over its reliability. It provides useful data for only 30%–40% of monitoring time because of possibility of monitor dysfunction due to placement errors.

Jugular venous oxygen saturation

Monitoring jugular venous oxygen saturation ($SjvO_2$) can be used to estimate the balance between global cerebral oxygen delivery and utilization. A catheter is inserted into the dominant (generally right side) internal jugular vein (IJV) and advanced to the jugular bulb, minimizing contamination from extracerebral venous return, which is around 3% if the catheter is correctly placed. Dominant side can be identified by the ultrasound examination of the IJV (whichever side has a larger diameter), larger ICP increase caused by manual compression of each IJV, and larger jugular foramen on the CT scan. Catheter positioning is checked on a lateral cervical spine radiograph and correct placement is identified when the tip of the catheter lies above the inferior border of C1 vertebra (Fig. 5.4). Measurement of $SjvO_2$ can be made either continuously using a fiber-optic catheter or directly by aspirating blood samples and using a cooximeter. $SJvO_2$ is dependent on technical aspects such as correct catheter placement to exclude the extracranial circulation. Global, flow-weighted measure is recorded, however it may miss critical regional ischemia. Values <50% have been associated with increased mortality and worse functional outcomes, as have the number of desaturation episodes experienced by a given patient. Upper limits >75% have also been recommended as a treatment threshold based on worse outcomes than in those with $SjvO_2$, 50%–74%. Arteriovenous difference in oxygen content ($AVDO_2$) has been advocated as a more relevant treatment value and is calculated using SaO_2, $SjvO_2$, and hemoglobin values and targeted values generally between 5% and 7.5%. Higher mean $AVDO_2$ values have been associated with favorable outcomes at 6 months. Positron emission tomographic evidence suggests that jugular SvO_2 does not decrease by <50% until 13% of the brain becomes ischemic.[40] Significant arteriovenous shunting, after TBI, might reduce the usefulness of this measurement. After early enthusiasm, its clinical use has decreased in favor of other methods of monitoring brain tissue oxygenation.

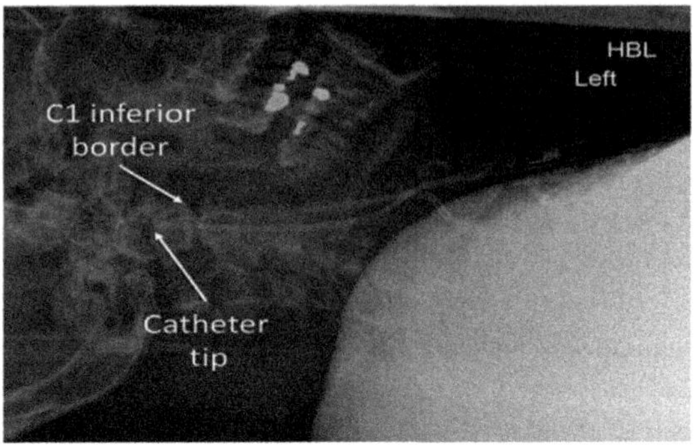

Figure 5.4 Jugular venous oxygen saturation catheter tip cephalad to inferior border of C1 in lateral radiographic image of neck.

Brain tissue oxygen partial pressure

It is the most promising of all bedside cerebral oxygenation monitoring techniques. The utility of the technique depends upon correct probe placement and knowledge of the location of the probe tip as it provides information about a small area of brain only. The partial pressure of oxygen in the cerebral interstitium is measured using optical luminescence (Neurovent-PTO system) or polarographic techniques (Licox system) depending on the type of microcatheter. The catheter is usually inserted into nontraumatized tissue which is identified on CT scan. If inserted into pericontusional tissue, $PbtO_2$ may produce lower readings than expected and underestimate the state of oxygen delivery to nontraumatized tissue. $PbtO_2$ is a complex variable influenced by *global determinants* of oxygen delivery such as PaO_2, $PaCO_2$, FiO_2, arterial blood pressure, cardiac output, hemoglobin, and cardiorespiratory function and *cerebral variables* including ICP, CPP, autoregulation, metabolism, seizures, and cerebral tissue oxygen gradients.

Normal $PbtO_2$ values range from 20 to 40 mmHg. Values less than 15–20 mmHg imply brain ischemia and values below 10 mmHg suggest severe ischemia. Brain tissue partial pressure of oxygen (PO_2) is best interpreted by taking in consideration both the duration and the severity of ischemia. Observational studies suggest better outcomes when ICP/CPP-guided management is supplemented with brain tissue PO_2-directed therapy to maintain brain tissue PO_2 more than 20 mmHg.[41] A systematic review of four studies incorporating 491 patients confirmed that brain tissue PO_2 and ICP/CPP-directed therapy combined is associated with superior outcomes compared to ICP/CPP-guided therapy alone.[42]

Okonkwo et al. carried out BOOST-II trial which was a two-arm, single-blind, prospective randomized controlled multicenter phase II trial assessing safety and efficacy of a management protocol optimizing $PbtO_2$ following severe TBI.[43] Patients were randomized to treatment protocol based on ICP plus $PbtO_2$ monitoring versus ICP monitoring alone. $PbtO_2$ data were recorded in the ICP-only group in blinded manner. Tiered interventions in each arm were specified and impact on

ICP and $PbtO_2$ measured. The authors concluded that management of severe TBI informed by multimodal ICP and $PbtO_2$ monitoring reduced brain tissue hypoxia with a trend toward lower mortality and more favorable outcomes than ICP-only treatment.

In general, data from $PbtO_2$ monitoring should be interpreted along with data from other monitoring parameters to make clinical decisions and should not be used in isolation. PbtO2 monitoring helps in determining the individual "optimal" CPP, that is, the CPP level to prevent brain hypoxia and monitor the effect of hemodynamic interventions. It is not only essential for optimization of CPP but also provides information to guide the interventions targeted to $PbtO_2$ and lactate:pyruvate ratio.

Near infrared spectroscopy

NIRS is a noninvasive technique. NIRS is based on the differential transmission and absorption of near infrared light (700–950 nm) as it passes through tissue. Oxygenated and deoxygenated hemoglobin have characteristic and different absorption spectra in the near infrared spectrum of light. Their relative concentrations in tissue can be determined by their absorption of light in this wavelength range. NIRS-based cerebral oximeters derive a scaled absolute hemoglobin concentration (the relative proportions of oxy- and deoxy-hemoglobin in the field of view) from which regional cerebral tissue oxygen saturation is calculated. It is sensitive to oxygen extraction and therefore provides regional assessment of critical oxygen supply/demand mismatch. The "normal" range of regional cerebral oxygen saturation lies between 60% and 75%, but there is substantial intra- and interindividual variability in NIRS-derived cerebral saturation. There are no validated regional cerebral saturation-defined ischemic thresholds to guide therapeutic interventions; small case series have demonstrated association of low regional cerebral oxygen saturation values with poor outcome, and also NIRS has been used to determine optimal CPP noninvasively.

However, data on the application of NIRS for monitoring after TBI are still limited, and no outcome studies investigating NIRS-guided management. NIRS has been used in monitoring of trauma patients preoperatively.[44] NIRS has been shown to allow early recognition of epidural and subdural hematomas in trauma cases.[45]

Brain metabolism and biochemistry

Microdialysis is an invasive technique used to detect secondary brain injury. The procedure consists of implantation of a probe in the region of the brain at risk. Microdialysis catheter has a semipermeable dialysis membrane at its tip. Diffusion drives the passage of molecules across the membrane along their concentration gradient from the brain extracellular fluid into the isotonic dialysis fluid. Microdialysis catheter is placed in at-risk brain tissue to monitor secondary insults in the area of brain most vulnerable to ischaemia (penumbra zone). Concentrations of clinically relevant compounds in the dialysate are measured in a semiautomated calorimetric bedside analyzer.

Imbalance in the supply and demand for glucose after brain injury can lead to a cerebral metabolic crisis from ischemic and nonischemic causes. Cerebral microdialysis is unique among bedside neuromonitoring techniques as it is able to identify both elevated lactate:pyruvate ratio with low pyruvate, which indicates a profound reduction in energy substrate supply and classic ischemia, as well as elevated lactate:pyruvate ratio with normal or high pyruvate which indicates a nonischemic cause related to mitochondrial dysfunction with or without increased metabolic demand.

This monitoring tool also monitors glutamate, which is a marker of hypoxia/ischemia and excitotoxicity, and glycerol which is a (nonspecific) marker of hypoxia/ischemia-related cell membrane breakdown. As microdialysis measures changes at the cellular level, it has the potential to identify cerebral compromise before changes in other monitored variables. Periods of low brain glucose concentration ($<0.7-1$ mmol/L) combined with elevated lactate:pyruvate ratio (>40) represent severe hypoxia/ischemia and are associated with poor outcomes.

Neuroglycopenia: After TBI, the normal relationship between serum glucose concentration, glycemic control, and brain glucose is disturbed, and brain glucose may fall to such low levels that are inadequate to meet the metabolic demand; although serum glucose concentration might be within a normal range. It indicates origin of a nonhypoxic metabolic crisis. If brain extracellular fluid glucose concentration is very low (0.2 mmol/L)—a trial of increasing serum glucose concentration (even if within normal limits) has been recommended to minimize the burden of neuroglycopenia.

The lactate:pyruvate ratio has been used to guide CPP management. Addition of CMD (in combination with $PbtO_2$) to ICP monitoring in patients with severe TBI markedly improves the accuracy of detecting secondary cerebral hypoperfusion. CMD markers—such as glucose and lactate:pyruvate ratio have been demonstrated to be good surrogate CBF markers. Studies combining microdiaylsis with brain imaging have confirmed this by following dynamic trends of both CMD lactate: pyruvate ratio and glucose detects secondary cerebral ischemic insults.[46,47]

Macromolecules can be sampled for research purposes using high-molecular-weight cut-off (100 kDa) dialysis membranes. Also, novel biomarkers, including S100B, nitric-oxide metabolites, and *N*-acetylaspartate, have been investigated. The incorporation of proteomics into the microdialysis technique has great potential.

The limitations of microdialysis include poor temporal resolution; hence, short-lived but important changes in brain tissue chemistry, including those induced by electrophysiologic abnormalities such as cortical spreading depolarizations may be missed. Also, there is "insertion effect," that is, the values obtained during the first 6 h after placement of the microdialysis probe can be unreliable.

In aneurysmal SAH patients, the use of cerebral microdialysis has been suggested to be useful in diagnosing vasospasm.[46] Post TBI, cerebral microdialysis has been used as a monitoring tool and has helped in prognostification of patients.[47]

Microdialysis after human occlusive stroke has shown massively elevated levels of glutamate, aspartate, lactate:pyruvate ratio, and glycerol.[48]

Electroencephalography

Electroencephalography (EEG) consists of recording of the summation of postsynaptic potentials spontaneously generated by the pyramidal cells. These signals are very small. Hence, extreme care must be taken when placing electrodes, either surface or subdermal needle electrodes. Seizures affect 20%–40% of TBI patients, and early detection and treatment will reduce the burden of seizure-related secondary injury due to increased expression of excitatory neurotransmitters and concomitant neuroinflammatory cascades. Continuous EEG monitoring is now recommended for the detection of posttraumatic seizures and to guide anticonvulsant therapy considering the high incidence of nonconvulsive seizures in patients with TBI. In 80% of TBI patients, EEG burst activity precedes an increase in ICP. Specific regions at risk of ischemic change can be identified by focal slowing of EEG. A single decrease in >50% alpha/delta wave ratio, specifically marks for the onset of ischemia. Intracranial EEG involves placement of electrodes in the subcortical region of the brain.

Invasive continuous EEG monitoring has been used to detect cortical spreading depolarizations, an important cause of secondary brain injury in more than half of TBI patients. Spreading depolarizations can currently only be detected by invasive EEG monitoring. Benefits of depth EEG include detecting many depth seizures which do not show up on a scalp EEG, and may predict a worse clinical outcome than cortical seizures alone, and depth EEG has a high signal-to-noise ratio and lesser patient artifact which is seen on surface EEG.

Postaneurysmal SAH, EEG monitoring helped detecting subclinical seizures and may predict delayed cerebral ischaemia (DCI) in advance.[49]

Evoked potentials

EPs are recordings of the electrical potentials produced after stimulation of specific sensory or motor tracts. Somatosensory EPs (SSEPs) and brainstem auditory-evoked responses (BAERs) are the two most commonly used types of EPs in the neurocritical care setting. They have been used as noninvasive diagnostic and prognostic tools, such as after cardiopulmonary resuscitation, after TBI, and also for brain death determination. BAER is sensitive to ponto-mesencephalic integrity, and early significant changes in V waves and V latency in case of trans-tentorial brain herniation and/or increased ICP have been demonstrated. EPs can be used to monitor brainstem compression in comatose patients.

Tibial nerve SSEP and motor EP are helpful biomarkers for prognostication of functional outcomes in patients with hemiplegia.[50,51] In patients with high-grade

SAH, median and tibial nerve SSEPs 2 weeks after aneurysmal hemorrhage has been suggested to predict long-term outcome.[52] There have been case reports of use of visual EPs in diagnosing persistently raised ICP.[53]

Conclusion

Individual modes of Multimodality neuromonitoring can help predict changes in cerebral and cerebrovascular physiology; however, integrated information from multiple modalities to reveal the patient-specific "injury profile" should be used to formulate an optimal treatment plan to improve patient outcomes rather than individual monitor values and treating individual thresholds. Early recognition of injury patterns using these monitors followed by timely administration of individually tailored treatments to prevent secondary brain damage should be the aim. However, till date, lots of research still needs to be done to extract meaningful information from these neuromonitoring devices in order to have robust guidelines for overall improvement of patient outcomes which have still not yet been achieved so far.

References

1. Teasdale G, Jennett B. Assessment of coma and impaired consciousness. A practical scale. *Lancet*. 1974;2:81–84.
2. Reith FC, Van den Brande R, Synnot A, Gruen R, Maas AI. The reliability of the Glasgow Coma Scale: a systematic review. *Intensive Care Med*. 2016;42(1):3–15.
3. Luk S, Jacobs L, Ciraulo D, Cortes V, Sable A, Dowell V. Outcome assessment of physiologic and clinical predictors of survival in patients after traumatic injury with a trauma score less than 5. *J Trauma*. 1999;46:122–127.
4. Wijdicks EF, Bamlet WR, Maramattom BV, Manno EM, McClelland RL. Validation of a new coma scale: the FOUR score. *Ann Neurol*. 2005;58(4):585–593.
5. Iyer VN, Mandrekar JN, Danielson RD, Zubkov AY, Elmer JL, Wijdicks EF. Validity of the FOUR score coma scale in the medical intensive care unit. *Mayo Clin Proc*. 2009;84(8):694–701.
6. Jamal A, Sankhyan N, Jayashree M, Singhi S, Singhi P. Full Outline of Unresponsiveness score and the Glasgow Coma Scale in prediction of paediatric coma. *World J Emerg Med*. 2017;8(1):55–60.
7. Idrovo L, Fuentes B, Medina J, et al. Validation of the FOUR score (Spanish version) in acute stroke: an interobserver variability study. *Eur Neurol*. 2010;63(6):364–369.
8. Eken C, Kartal M, Bacanli A, Eray O. Comparison of the Full Outline of Unresponsiveness score coma scale and the Glasgow Coma Scale in an emergency setting population. *Eur J Emerg Med*. 2009;16(1):29–36.
9. Suresh V, Yaddanapudi LN, Podder S. Full Outline of UnResponsiveness score vs Glasgow Coma Scale in critically ill patients with altered sensorium: a comparison of inter-observer variability and outcomes. *Indian J Anaesth*. 2019;63(8):640–647.
10. Almojuela A, Hasen M, Zeiler FA. The Full Outline of UnResponsiveness (FOUR) score and its use in outcome prediction: a scoping review of the pediatric literature. *J Child Neurol*. 2019;34(4):189–198.

11. Hunt WE, Hess RM. Surgical risk as related to time of intervention in the repair of intracranial aneurysms. *J Neurosurg*. 1968;28(1):14–20.
12. World Federation of Neurological Surgeons Committee. Universal subarachnoid hemorrhage grading scale. *J Neurosurg*. 1988;68(6):985–986.
13. Fisher CM, Kistler JP, Davis JM. Relation of cerebral vasospasm to sub-arachnoid hemorrhage visualized by computerized tomographic scanning. *Neurosurgery*. 1980;6: 1–9.
14. Frontera JA, Claassen J, Schmidt JM, et al. Prediction of symptomatic vasospasm after subarachnoid haemorrhage: the modified fisher scale. *Neurosurgery*. 2006;59(1):21–27.
15. Fujimoto K, Miura M, Otsuka T, Kuratsu J. Sequential changes in Rotterdam CT scores related to outcomes for patients with traumatic brain injury who undergo decompressive craniectomy. *J Neurosurg*. 2016;124(6):1640–1645.
16. Katar S, Aydin OP, Ozel M, et al. The use of Rotterdam CT score for prediction of outcomes in pediatric traumatic brain injury patients admitted to emergency service. *Pediatr Neurosurg*. 2020;55(5):237–243.
17. Bae IS, Chun HJ, Yi HJ, Choi KS. Using components of the Glasgow coma scale and Rotterdam CT scores for mortality risk stratification in adult patients with traumatic brain injury: A preliminary study. *Clin Neurol Neurosurg*. 2020;188:1055–1059.
18. Zacchetti L, Magnoni S, Di Corte F, Zanier ER, Stocchetti N. Accuracy of intracranial pressure monitoring: systematic review and *meta*-analysis. *Crit Care*. 2015;2(19):420.
19. Rao SS, Chung DY, Wolcott Z, et al. Intermittent CSF drainage and rapid EVD weaning approach after subarachnoid haemorrhage: association with fewer VP shunts and shorter length of stay. *J Neurosurg*. 2019;26:1–6.
20. Bales JW, Bonow RH, Buckley RT, Barber J, Temkin N, Chesnut RM. Primary external ventricular drainage catheter vs intraparenchymal ICP monitoring: outcome analysis. *Neurocrit Care*. 2019;31(1):11–21.
21. Koskinen LO, Grayson D, Olivecrona M. The complications and the position of the Codman MicroSensor™ ICP device: an analysis of 549 patients and 650 sensors. *Acta Neurochir (Wien)*. 2013;155(11):2141–2148.
22. Chambers IR, Siddique MS, Banister K, Mendelow AD. Clinical comparison of the Spiegelberg parenchymal transducer and ventricular fluid pressure. *J Neurol Neurosurg Psychiatry*. 2001;71(3):383–385.
23. Chesnut RM, Temkin N, Carney N, et al. A trial of intracranial-pressure monitoring in traumatic brain injury. *N Engl J Med*. 2012;367(26):2471–2481.
24. Myers RB, Lazaridis C, Jermaine CM, Robertson CS, Rusin CG. Predicting intracranial pressure and brain tissue oxygen crises in patients with severe traumatic brain injury. *Crit Care Med*. 2016;44(9):1754–1761.
25. Blaivas M, Theodoro D, Sierzenski PR. Elevated intracranial pressure detected by bedside emergency ultrasonography of the optic nerve sheath. *Acad Emerg Med*. 2003; 10(4):376–378.
26. Dubourg J, Javouhey E, Geeraerts T, Messerer M, Kassai B. Ultrasonography of optic nerve sheath diameter for detection of raised intracranial pressure: a systematic review and *meta*-analysis. *Intensive Care Med*. 2011;37(7):1059–1068.
27. Newman WD, Hollman AS, Dutton GN, Carachi R. Measurement of optic nerve sheath diameter by ultrasound: a means of detecting acute raised intracranial pressure in hydrocephalus. *Br J Ophthalmol*. 2002;86(10):1109–1113.
28. Mostafa AM, Derakhshanfar H, Noori S, et al. Agreement between measuring optic nerve sheath diameter by ultrasonography and brain CT scan among patients with traumatic brain injury. *Int J Surg Med*. 2018;4(3):123–127.

29. Keyes LE, Paterson R, Boatright D, Browne V, Leadbetter G, Hackett P. Optic nerve sheath diameter and acute mountain sickness. *Wilderness Env Med*. 2013;24(2):105–111.
30. Bloria SD, Bloria P, Luthra A. Is it the time to standardize the procedure of ultrasound guided optic nerve sheath diameter measurement? *Saudi J Anaesth*. 2019;13(3): 255–256.
31. Aaslid R, Lindegaard KF, Sorteberg W. Cerebral autoregulation in humans. *Stroke*. 1989;20:45.
32. Zanette EM, Fieschi C, Bozzao L. Comparison of cerebral angiography and transcranial Doppler sonography in acute stroke. *Stroke*. 1989;20:899.
33. Demarin V, Rundek T, Hodek B. Maternal cerebral circulation in normal and abnormal pregnancies. *Acta Obstet Gynecol Scand*. 1997;76:619.
34. Strauss G, Hansen BA, Knudsen GM. Hyperventilation restores cerebral blood flow autoregulation in patients with acute liver failure. *J Hepatol*. 1998;28:199.
35. Riker RR, Sawyer ME, Fischman VG, et al. Neurological pupil index and pupillary light reflex by pupillometry predict outcome early after cardiac arrest. *Neurocrit Care*. 2020;32(1):152–161.
36. Miroz JP, Ben-Hamouda N, Bernini A, et al. Neurological pupil index for early prognostication after venoarterial extracorporeal membrane oxygenation. *Chest*. 2020;157 (5):1167–1174.
37. Jahns FP, Miroz JP, Messerer M, et al. Quantitative pupillometry for the monitoring of intracranial hypertension in patients with severe traumatic brain injury. *Crit Care*. 2019;23(1):155.
38. Freeman AD, McCracken CE, Stockwell JA. Automated pupillary measurements inversely correlate with increased intracranial pressure in pediatric patients with acute brain injury or encephalopathy. *Pediatr Crit Care Med*. 2020;21(8):753–759.
39. Osman M, Stutzman SE, Atem F, et al. Correlation of objective pupillometry to midline shift in acute stroke patients. *J Stroke Cerebrovasc Dis*. 2019;28(7):1902–1910.
40. Coles JP, Fryer TD, Smielewski P, et al. Incidence and mechanisms of cerebral ischemia in early clinical head injury. *J Cereb Blood Flow Metab*. 2004;24:202–211.
41. Lin CM, Lin MC, Huang SJ, et al. A prospective randomized study of brain tissue oxygen pressure-guided management in moderate and severe traumatic brain injury patients. *Biomed Res Int*. 2015;2015:529–580.
42. Nangunoori R, Maloney-Wilensky E, Stiefel M, et al. Brain tissue oxygen-based therapy and outcome after severe traumatic brain injury: a systematic literature review. *Neurocrit Care*. 2012;17(1):131–138.
43. Okonkwo DO, Shutter LA, Moore C, et al. Brain oxygen optimization in severe traumatic brain injury phase-II: a phase II randomized trial. *Crit Care Med*. 2017;45 (11):1907–1914.
44. Beekley AC, Martin MJ, Nelson T, et al. Continuous noninvasive tissue oximetry in the early evaluation of the combat casualty: a prospective study. *J Trauma*. 2010;69(Suppl 1):S14–S25.
45. Kessel B, Jeroukhimov I, Ashkenazi I, et al. Early detection of life-threatening intracranial haemorrhage using a portable near-infrared spectroscopy device. *Injury*. 2007;38 (9):1065–1068.
46. Unterberg AW, Sakowitz OW, Sarrafzadeh AS, Benndorf G, Lanksch WR. Role of bedside microdialysis in the diagnosis of cerebral vasospasm following aneurysmal subarachnoid hemorrhage. *J Neurosurg*. 2001;94(5):740–749.
47. Goodman JC, Valadka AB, Gopinath SP. Extracellular lactate and glucose alterations in the brain after head injury measured by microdialysis. *Crit Care Med*. 1999;27:1965–1973.

48. Berger C, Annecke A, Aschoff A. Neurochemical monitoring of fatal middle cerebral artery infarction. *Stroke*. 1999;30:460–463.
49. Kondziella D, Friberg CK, Wellwood I, et al. monitoring in aneurysmal subarachnoid hemorrhage: a systematic review. *Neurocrit Care*. 2015;22(3):450–461.
50. Hwang P, Sohn MK, Kim CS, Jee S. Tibial somatosensory evoked potential can prognosticate for ambulatory function in subacute hemiplegic stroke. *J Clin Neurosci*. 2016;26:122–125.
51. Hwang P, Sohn MK, Jee S, Lee H. Transcranial motor evoked potentials of lower limbs can prognosticate ambulation in hemiplegic stroke patients. *Ann Rehabil Med*. 2016;40 (3):383–391.
52. Mende KC, Gelderblom M, Schwarz C, et al. Somatosensory evoked potentials in patients with high-grade aneurysmal subarachnoid hemorrhage. *Neurosurg Focus*. 2017;43(5):E17.
53. Liasis A, Thompson DA, Hayward R, Nischal KK. Sustained raised intracranial pressure implicated only by pattern reversal visual evoked potentials after cranial vault expansion surgery. *PediatrNeurosurg*. 2003;39(2):75–80.

Cognition and translational research

6

Francesca Rubulotta[1], *Ahmed Ezzat*[2,3], *Zena Radi*[4] *and*
Federico Billotta[5]
[1]Chair of the Department of Critical Care, McGill University, School of Medicine and Health Science, Montreal, QC, Canada, [2]Department of Surgery and Cancer, Imperial College Healthcare NHS Trust, London, United Kingdom, [3]Department of Surgery and Cancer, Faculty of Medicine, Imperial College London, London, United Kingdom, [4]Sapienza University of Rome, Rome, Italy, [5]Anaesthesia and Intensive Care Medicine, Sapienza University of Rome, Rome, Italy

Introduction

The word cognition dates back to the 15th century and it meant "thinking and awareness." The cognitive process has been studied for more than 18 centuries, beginning with Aristotle (384–322 BC) who described cognitive areas connected to memory, perception, imagination, and creativity.[1] More recently, cognition refers to "the mental action or process of acquiring knowledge and understanding through thoughts, experience, and perceptions." Research into cognitive impairments started 100 years ago, but findings related to medicine and surgery were enhanced by the development of the neuropsychological testing at the end of last century.[1] The translational research related to cognition incorporates aspects of both basic science and clinical research. Translational research implements a concept from "bench-to-bedside," from laboratory experiments through clinical trials to point-of-care. The end point of translational research is the production of a promising new treatment that can be used at the bedside. This chapter will focus on translational research and cognition. This manuscript is intended to summarize examples of translational research with particular focus on cognitive dysfunction during the perioperative period. Risk factors that could lead to cognitive impairments in this population include age, prolonged hospitalization, sleep deprivation, pain, anxiety, fitness, specific surgical challenges such as bleeding, anesthesia, and invasive mechanical ventilation (MV). There are significant limitations related to the use of translational research, these are mainly due to inconsistencies in several preclinical trials.

Cognitive dysfunctions in the perioperative period: definition and diagnosis

Perioperative medicine is a new multidisciplinary subspeciality addressing the complex needs of patients undergoing surgery. Perioperative medicine encompasses the

Perioperative Neuroscience. DOI: https://doi.org/10.1016/B978-0-323-91003-3.00009-X

care of patients from the time of preparing for surgery, through the operative period to full recovery. Patients are treated in the perioperative period by surgeons, anesthesiologists, intensivists, internal medicine generalists, pulmonologists, cardiologists, physiotherapists, nutritionists, diabetologists, nephrologists, psychotherapists, and many more. The medical knowledge required includes patient's specific risks, surgical specific risks, and the management of complex medical illness or comorbidities. The definition, classifications, and research related to perioperative cognitive changes have been done in segregation from investigations in the general population. As a matter of fact, the two major classification guidelines [Diagnostic and Statistical Manual for Mental Disorders, fifth edition (DSM-5) and National Institute for Aging and the Alzheimer Association] are used outside the fields of anesthesia and surgery. Originally, the postoperative cognitive dysfunction (POCD) was defined by any "deterioration in cognitive performance."[2] This condition has been initially described in elderly patients—aged ≥65 years old—and cognitive changes have been recorded up to 7.5 years after surgery.[2,3] In 2018, an international working group has determined that "perioperative neurocognitive disorders" (PNDs) include

- "Cognitive decline" diagnosed before operation (neurocognitive disorder);
- Any form of "acute event" [including postoperative delirium (POD)];
- "Cognitive decline diagnosed up to 30 days after the procedure (delayed neurocognitive recovery) and up to 12 months (postoperative neurocognitive disorder)."[2]

The diagnosis of POCD is frequently missed or delayed because of the lack of a consistent approach toward defining POCD.[2] A recent focused systematic review has included 30 studies to measure the degree of heterogeneity in the definition of POCD.[4] A neurocognitive test was performed according to this review shortly before and one week after surgery. A variety of neurocognitive tests were used including complex attention, language, executive functioning, perceptual-motor function, learning, and memory.[4] Authors adopted most commonly the Mini–Mental State Examination, the digit span test, the trail making test part A, and the digit symbol substitution test.[2,4] Results from this systematic review combined with conclusions from the Nomenclature Consensus Working Group experts' report in 2018 failed to identify a consensus on which test is the gold standard for the diagnose of POCD.[2,4] Therefore authors of the systematic review concluded that there is a lack of a consistent approach toward defining POCD.[4] Nevertheless, the formal classification presented in 2018 has been crucial given that the incidence of POCD might have been higher than reported for decades and that this number will increase over time.[2] As a matter of fact, the number of patients undergoing surgery under general anesthesia aged >60 years has significantly increased. The estimates suggest that 25% of all surgeries will be performed on >60-year-old people by 2025.[2] The terminology related to POCD is aligned to the DSM-5 since 2018 and this might enhance cross specialty research and improvements in the clinical recognition and management of these disorders. On the other hand, neurocognitive disorders are frequent in the community. A range between 14% and 48% of people experience neurocognitive disorders at the age more than 70 years and 10% suffer of dementia. This has an impact on outcome following surgery under general anesthesia. The reflection on society and costs is immense because several of these patients are hospitalized for long-time or even for the rest of their life after surgery. These individuals have cognitive disorders before surgery and this element is vastly missed in the preoperative assessment. The neurocognitive pathologies described in the general population were defined using a different terminology compared to the perioperative disorders. Currently, definitions are similar and the treatments are correlating in the attempt of improving clinical outcome.

In 1955 Bedford et al. published a case series involving 18 mentally normal people aged 67–97-year-old undergoing surgery under general anesthesia.[5] Authors noted that these patients were confused, torpid, unaware of their surroundings, unable to recognize their relatives, unable to converse, able to walk only with assistance, and some could not even remember their name.[5] In response to this study patients' age, surgery and general anesthesia were highlighted as the three main risk factors associated with the onset of POCD.[5–8] POD was also frequently diagnosed in hospitalized patients up to 1 week postsurgery and it was associated with poor outcome in the elderly.[9] The most recent definition of POD and POCD suggests that these are comprised in the term PND.[10] According to the definition, POD is considered as an acute event, whereas POCDs in general are mild or major neurocognitive disorders that can occur from 1 to 12 months postoperation. POD is characterized by fluctuating disturbance in attention and altered awareness, while POCDs are associated with a broadly reduced ability of learning and memory.

Translational research in PND aims at identifying possible underlying causative mechanisms and risk factors using animal models.[11,12] Translational research turns preclinical findings on cognitive decline into effective diagnostic or therapeutic tactics.[13–17] Various aspects from the whole perioperative period, which includes preoperative, intraoperative, and postoperative (intensive care admission and/or rehabilitation), have been tested bench-to-bedside. This research has confirmed the initial theory, that is age, surgery, and anesthesia are related to PND.[10–14] Nevertheless, length of hospital stay, sleep deprivation, preoperative and postoperative pain, anxiety, muscle injury, preexisting cognitive impairment, sensory impairment, and the use of MV during general anesthesia or in the intensive care unit (ICU) after surgery have also been identified as independent rick factors for PND.[17–21]

Authors will summarize animal models of cognitive dysfunction which have been crucial to advance diagnosis and treatment of PND. Perioperative cognitive evaluation is difficult to study using animal models but some experiments have been crucial to guide physicians at the bedside.

Animal models of cognitive dysfunction in perioperative neuroscience

The impact of general anesthesia combined with prolonged invasive MV is significant and these can trigger neuroinflammation events as well as memory loss in male mice.[22] Chen et al. randomly divided male mice into three groups: a control group where mice on spontaneous breathing mode did not receive surgery but only analgo-sedation; a surgery group where animals on spontaneous breathing mode underwent orthopedic surgery under general anesthesia; and a MV group where mice under invasive MV modes for 1, 3, or 6 hours received general anesthesia and the same orthopedic surgery as the second group.[22] Mice were tested for memory loss using "fear-conditioning" tests. Blood was collected as well as hippocampus and brain tissue samples. It was observed that mice in the MV group for more than 6 hours had increased CD11b-immunopositive cells, synapse degeneration, cytochrome c release, cleaved caspase-3, and cleaved poly(ADP-ribose) polymerase-1 activation. Mice receiving general anesthesia on spontaneous breathing mode did not show similar changes. Prolonged invasive MV also increased nuclear factor kappa B activation and levels of systemic and hippocampal interleukin (IL)-1β, IL-6, and tumor necrosis factor (TNF)-α after surgery. This study showed that MV for longer than 6 h may aggravate cognitive decline after surgery. General anesthesia and sedation using spontaneous breathing modes did not show similar changes. Prolonged invasive MV following surgery and/or due to postoperative critical illness caused also significant long-term neurological morbidity.[23]

Further experimental animal studies have been done to explore possible mechanisms related to pathways connecting lungs to brain degeneration during invasive MV. Protective ventilation strategies are used to prevent adverse effects in MV patients.[19] These ventilation strategies minimize lung stretch which results in preventing failure of distal organs, including the brain. The connection between lung and brain, involving specific pathways during critical illness, deserves further efforts to evaluate, prevent and improve cognitive alterations after long-term invasive MV. It has been observed that alveolar stretching, $PaO_2/PaCO_2$ abnormalities, and cytokine and chemokine expression in the lungs are the main mechanisms that possibly contribute to MV-induced brain damage.[21] Alveolar stretching promotes the production of systemic and cerebral mediators that lead to neuroinflammation, dopaminergic neuronal signaling, and beta amyloid synthesis.[21] The $PaO_2/PaCO_2$ abnormalities, particularly hypoxia and hypercapnia, might contribute to promote neuronal death, cerebral oxidative stress, and changes in brain blood flow that worsen brain damage. Prolonged invasive MV triggers brain inflammation inducing cerebral expression of proinflammatory cytokines (TNF-α, IL-1β, and IL-6) and microglia activation. Increased systemic inflammatory mediators produced by the lungs might directly change neurocognitive functions. Bilotta et al. recommend strategies to monitor and prevent MV-induced brain damage and the related neurocognitive decline caused by neuroinflammatory scavengers, immune modulation, DRD2-blockers, and lung "neuroprotective" MV management could be effective strategies.[21]

ICU-related neurocognitive impairments are even more severe compared to perioperative cognitive sequelae.[23] ICU survivors experience memory loss, attentional gaps, and slower processing speed.[23] These sequelae have an important impact on patients' lives, and ICU survivors often require institutionalization and further prolonged hospitalization. Hospitalization is also an independent risk factor for PND which will be examined later in this chapter. Long-term neurocognitive impairments experienced by critically ill survivors may be mitigated by early interventions, combining cognitive and physical therapies.[19] Inpatient rehabilitation interventions in ICU promise to improve outcomes in critically ill patients.[19] Prolonged postoperative ICU stay correlates with neuropsychological disturbances including acute delirium or memory impairment mostly after ICU discharge.[23] Therefore patients undergoing invasive MV for long periods show neurologic impairment with memory and cognitive alteration similar to the experiment performed using male mice. Delirium cannot easily be studied using animal models like memory or general cognitive dysfunction. This is unfortunate given that delirium duration is an independent predictor of mortality, ventilation time, ICU length of stay, and short- and long-term cognitive impairments.[23,24]

Anesthetics

Existing literature continues to debate whether short-term effects on cognition can be attributed to anesthetic agents or whether other variables are causative. Furthermore, there is conflicting data on the relationship between anesthesia and the development of long-term neurocognitive disorders or development of incident dementia. The type of anesthetic agent used, mode of administration, and depth of anesthesia play a role in PND.[24,25] POD has several implications such as increased hospital stay, mortality, and cognitive decline leading to functional dependence requiring higher rates of nursing home requirements.[26] During early postnatal life, animal and epidemiologic studies have also raised a concern that exposure to anesthetic agents may cause lasting impairments in cognitive function. Ongoing research, particularly in rodents, has addressed this question too guiding pediatric guidelines.[25] Xie et al. published a laboratory study in 2006 where the inhalational anesthetic agent isoflurane was exposed at a concentration of 2% for 6 hours to naïve H4 human neuroglioma cells (H4-APP-FL cells), and effects compared against a control consisting of 2% CO_2 and 21% O_2 and balanced nitrogen.[26] Western blot analysis was used in caspase-3 activation quantification (a late marker of cell apoptosis). The results demonstrated, the cells exposed to 2% isoflurane showed a 250% rise in the quantity of caspase-3, in comparison with the cells exposed to the control. Isoflurane demonstrated further reduction of cell viability by a quarter in comparison to the control. The study also reports the anesthetic agent impact amyloid precursor protein processing and amyloid beta production. Levels of amyloid following 2% isoflurane exposure demonstrated a three to four-fold increase in comparison to the control conditions (193 vs 50 pg/mL), respectively. Amyloid beta is typically located within the brain and is often linked with advancing age and is considered as a major cause underlying the pathophysiology of Alzheimer's disease (AD).[27]

A French study by Le Freche et al. studied 6-month-old mice to evaluate the effects of 1.5% or 2.5% inhalational anesthetic sevoflurane exposure to POCD. The study exposed the mice to 1 hour of the anesthetic (group 1) versus repeated exposures of five times for 1 hour in a month (group 2). The author reports exposure to the anesthetic led to statistically significant and dose related hippocampal tau phosphorylation which was reversible in group 1. In group 2 mice cognitive testing demonstrated persistent cognitive impairment.[28] Tau proteins are critical in the formation of microtubules which constitute part of the neuronal axon, enabling its function. Tau proteins also play a role within the neuronal cells and dendrites.[29] Phosphorylation of particular tau proteins abnormally may result in defective neuronal formation, observed in neurodegenerative illnesses such as AD and memory impairment. The literature reports anesthetic exposure during surgery may suppress neuronal activity sufficiently to result in POCD. During general anesthesia with volatile agents intraoperative electroencephalography (EEG) exhibits dominant delta waves (0–4 Hz) and concomitant theta waves (4–8 Hz) with or without alpha waves (8–12 Hz) and beta waves (12–16 Hz).[30] The presence of low frequency EEG waveforms poses an increased risk to POD, as demonstrated is cardiac patients postoperatively.[31] Increased anesthetic agent circulating in sedated patients can cause suppression of EEG bursts, defined as suppressed periods interrupted by brief high amplitude waveform bursts.[32] EEG burst suppression is pathological in the absence of general anesthesia and does not occur in natural human sleep. It has been linked with pathological processes and poor patient outcomes like delirium, cognitive impairment, and 1-month mortality in mechanically ventilated patients.[33]

Fritz et al. conducted a study in 2016 which included 730 patients undergoing surgery under general anesthetic, requiring a planned postoperative intensive care stay. Patient neuronal activity was measured using real time intraoperative EEG whilst anesthetized.[33] POD was measured on day 1–5 postoperatively by the administration of the Confusion Assessment Method. A 30 days postoperatively cognition and level of functional independence were measured using the PROMIS Applied Cognition-Abilities Short Form survey. Overall, nearly a quarter of patients developed POD. Patients who underwent longer periods of intraoperative EEG suppression demonstrated increased incidence of POD. At 30-day follow up postoperatively, patients with intraoperative EEG suppression showed worse scores measuring functional independence. Similar to other literature, the study reports higher concentrations of anesthetic agents and inadequate opiate use risked EEG suppression. EEG burst suppression may reflect high depth of anesthesia causing increased cortical stimulation and depletion of important calcium stores within the extracellular spaces cascading neuronal function impairment resulting in cognitive decline postoperatively.[34] Another mechanism elucidated in the literature is subclinical cognitive vulnerability in some patients, for whom anesthesia strains their "vulnerable" cognitive reserve edging them toward POCD. This is especially relevant to those patients who have underlying pathophysiological disease, amyloid and tau deposition but yet are subclinical and will remain cognitively competent for many decades.[35] EEG suppression may not only reflects increased depth of anesthesia but cerebral hypoperfusion resulting from shock such as hemorrhagic or anesthesia-related hypotension. As such current trials such as the SHARP trial (shaping anesthetic techniques to reduce postoperative delirium) are studying the benefits of standard intraoperative EEG use to measure depth of general anesthesia versus targeted spinal anesthesia in spinal surgery to evaluate short- and long-term outcomes on delirium, cognitive function, and physical independence.[36] On the contrary, a 2017 systematic review comparing regional versus general anesthetic on POD in orthopedic patients with hip fractures identified 100 studies. They found no link between the type of anesthesia and POD, although showed increased hospital stay and intraoperative hypotension and respiratory complications in general anesthesia.[37] Clearly, this is an area of academic debate, and future research will undoubtedly elucidate these questions. Anesthetic gases induce changes which may reproduce AD pathology in mice with mutations which produced AD. It would be interesting to know whether anesthetics are risky for subjects with special genetic risk factors.

Clinical studies investigating cognitive dysfunction in perioperative neuroscience

Hospitalization

The role of hospitalization as a possible risk factor for PND was analyzed in a multicenter study that enrolled 372 patients aged ≥ 60 years undergoing minor surgery.[38] About 199 patients were admitted for inpatient care while 173 were treated as out-patients. The cognitive function was assessed using the activity of daily living (IADL) questionnaire preoperatively, 7 days and 3 months postoperatively.[38] The IADL comprises seven questions related to walking, handling money, preparing meals, taking drugs/medicine, shopping, housework, and using a telephone. The IADL questionnaire was answered by both patients and relatives. The incidence of PND was higher in the inpatient group compared to the out-patients (odds ratio = 2.8 [1.05–7.4], $P = .04$). The patients who underwent out-patient surgery had significantly better IADL scores at the preoperative and 7-day postoperative evaluations. Risk factors for PND especially in elderly patients in this study included isolation, anxiety, noise, bright light, immobilization, and sleep deprivation. The authors concluded suggesting avoiding hospitalization of older patients when possible.[38]

Surgery including bleeding requiring blood transfusions

Perioperative blood loss and consequent administration of blood transfusion for patients have been linked to PND, specifically POD.[39] One study described the relationship between intraoperative blood transfusion and postoperative systemic inflammatory response syndrome (SIRS). The study included more than 500,000 adult patients who underwent surgery between 2005 and 2009. Following propensity matching the authors found a twofold increase in the rate of SIRS linked to intraoperative transfusion.[39] SIRS is widely reported for its role in sepsis-associated delirium.[40] Notably, the development of postoperative SIRS increases the risks for infections, renal insufficiency, strokes, and myocardial events, all of which are linked with cognitive dysfunction in surgery.[39] Often, the patient demographic is one which is elderly and comorbid, complicated by poor physiological reserve. Transfusion related SIRS is thought to involve activation of the humoral immune reaction via phagocytes and systematic cytokine release. Data suggest stored blood products contain activated proinflammatory mediators such as circulating TNF-α, IL-B, 6 and 8, all of which are associated with POCD.[41,42] Proinflammatory cytokine release may result in delirium via transmission to the brain via the blood—brain barrier and circumventricular areas. Consequently, microglial activation occurs and further cytokines are triggered within cerebral tissues impacting on overall cognitive function causing impairment.[43] Proinflammatory cytokines are widely accepted as a cause of delirium and are hypothesized to inhibit the production of acetylcholine via reducing immune-reactive neurons containing choline acetyltransferase.[44]

One trial studied more than 300 adult patients undergoing elective total hip replacements. Authors analyzed independent risk factors associated with POD. The study assessed cognitive status by administration of a validated cognitive function test preoperatively and at 7 days postoperatively. Of the 27% of patients who developed POCD at 7 days statistical significance was noted with increase intraoperative blood loss, lower hemoglobin preoperatively, and higher volume of units of blood transfusion peroperatively.[45] Behrends et al. conducted a similar study whereby 472 patients undergoing major noncardiac surgery were included.[46] Patients were 65 years or older and underwent knee, hip, spinal, abdominal, and thoracic operations. Patients underwent structured interviews preoperatively and on day 1 and 2 postoperatively. Authors used the Confusion Assessment Method.[46,47] Behrends et al. reports 29% of the participants went on to develop delirium defined as the acute onset and fluctuating course of mental status impairment, including inattention and disorganized thinking or altered state of consciousness.[46,48] Patients who suffered delirium were associated with receiving blood transfusions (40.3% vs 25.8%) and also more blood volume (850 ± 1191 mL vs 617 ± 1224 mL) in comparison with patients who did not require any blood products. Preoperative anemia was associated with POD. Receiving more than 1000 mL of red blood cells was the most significant predictor of early POD on the first day postoperatively (odds ratio = 3.68, 95% confidence interval = 1.32–10.94; $P < .01$). Another study, which included more than 1300 patients, identified intraoperative blood loss and the total number of blood units transfused were linked with POD.[49,50]

Early recovery after surgery (ERAS) pathways had a slow implementation over the last decades. ERAS pathways have defined recovery based on morbidity, length of hospital stay, and readmission rate, which do not indicate "recovery" from the patient's perspective. Although preliminary observations on ERAS are successful, the overall results in elderly patients have not been convincing. Future challenges for perioperative care and rehabilitation should include a more patient centered approach.[51]

Perioperative care for patients with known presurgery cognition dysfunctions

Patients with preexisting cognitive impairments present a unique set of considerations, including challenges with cooperation during the preoperative assessment, during the procedure if performed under sedation, or regional anesthesia techniques and with general anesthetic drug interactions. A recent review highlights clinical and preclinical evidence which implicates neuroinflammation, calcium dysregulation, and mitochondrial dysfunction.[52] Recent clinical studies have suggested that there is an increased risk of AD in patients undergoing surgical interventions, but it is unknown whether this effect is related to anesthesia, cardiovascular complications of surgery, or associated conditions such as hypothermia. AD and dementia are frequent preoperatively in the elderly and have an impact on PND. Exposure of old rodents to anesthesia produced memory deficits and increased levels of amyloid beta peptide and phosphorylated tau in brain. The effects of long- or short-term repetitive exposure to small molecular weight anesthetics are more severe in transgenic AβPPswe than in wild type mice. The most sensible approach is to refer patients >60 years of age for neuropsychometric testing to assess preoperative cognitive function. Experts in the field of anesthesia advice specific adjustments in the perioperative care, preparation for surgery, anesthesia techniques, and postoperative recovery to reduce the incidence of POCD in high-risk individuals or those with preexisting conditions. It is often the patient and/or a relative that initially detects a change in cognitive capacity after surgery, typically when resuming daily activities.

Limits of translational research for inconsistencies of evidence from preclinical studies

Animal models are essential to support translational research. Nevertheless, predictive validity of these models is important, the study design might be challenging and broad results might be difficult to apply to human research or models. Reversely, clinical bedside findings that were not predicted by animal testing should be back translated and used to refine the animal models.[53] Applied animal models need standardization and validation and this should be done by experts. The purpose is finding reproducible criteria which produce results applicable to humans and vice versa.

Conclusion

POCD is any decline in cognition present in the perioperative period including when a patient has recovered from the acute impact of surgery, ICU, and hospital stay. Since 2018, POCD can be defined as "delayed neurocognitive recovery" if present within 30 days of surgery, or “postoperative neurocognitive disorder” if present within 12 months after surgery. The lack of gold standard in testing for POCD and POD has led to significant delays in the diagnosis and treatment of these patients. There are several risks factors that could be identified and managed to reduce the risk of developing POCD. Animal studies can guide research in this field. Elderly patients might suffer cognitive impairments including AD and dementia preoperatively. These conditions are challenging and might lead to prolonged hospitalization and generalized deterioration of these patients. ERAS and novel perioperative measures are needed to prevent POCD and POD.

References

1. Franchi S, Bianchini F. *On the historical dynamics of cognitive science: a view from the periphery. The Search for a Theory of Cognition: Early Mechanisms and New Ideas.* Amsterdam, New York: Rodopi; 2011:xi−xxvii. Available from: http://hdl.handle.net/11585/108063.
2. Evered L, Silbert B, Knopman DS, et al. Recommendations for the nomenclature of cognitive change associated with anaesthesia and surgery. *Can J Anaesth.* 2018;65(11):1248−1257. Available from: https://doi.org/10.1007/s12630-018-1216-x.
3. Savage GH. Insanity following the use of anæsthetics in operations. *Br Med J.* 1887;2 (1405):1199−1200.
4. van Sinderen K, Schwarte LA, Schober P. Diagnostic criteria of postoperative cognitive dysfunction a focused systematic review. *Anesthesiol Res Pract.* 2020;2020:7384394. Available from: https://doi.org/10.1155/2020/7384394. Published online November 16, 2020.
5. Simpson BR, Williams M, Scott JF, Smith AC. The effects of anesthesia and elective surgery on old people. *Lancet.* 1961;2(7208):887−893. Available from: https://doi.org/10.1016/S0140-6736(61)90403-2.
6. Nielson WR, Gelb AW, Casey JE, Penny FJ, Merchant RN, Manninen PH. Long-term cognitive and social sequelae of general vs regional anesthesia during arthroplasty in the elderly. *Anesthesiology.* 1990;73(6):1103−1109. Available from: https://doi.org/10.1097/00000542-199012000-00006.
7. O’Keeffe ST, Chonchubhair AN. Postoperative delirium in the elderly. *Br J Anaesth.* 1994;73(5):673−687. Available from: https://doi.org/10.1093/bja/73.5.673.
8. Moller JT, Cluitmans P, Rasmussen LS, et al. Long-term postoperative cognitive dysfunction in the elderly ISPOCD1 study. ISPOCD investigators. International Study of Post-Operative Cognitive Dysfunction. *Lancet.* 1998;351(9106):857−861. Available from: https://doi.org/10.1016/s0140-6736(97)07382-0.
9. Deng Y, Wang R, Li S, et al. Methylene blue reduces incidence of early postoperative cognitive disorders in elderly patients undergoing major non-cardiac surgery: an open-label randomized controlled clinical trial. *J Clin Anesth.* 2020;68:110108. Available from: https://doi.org/10.1016/j.jclinane.2020.110108.
10. Evered L, Silbert B, Knopman DS, et al. Recommendations for the nomenclature of cognitive change associated with anaesthesia and surgery-2018. *Br J Anaesth.* 2018;129(5):872−879. Available from: https://doi.org/10.1097/ALN.0000000000002334.
11. Eckenhoff RG, Maze M, Xie Z, et al. Perioperative neurocognitive disorder: state of the preclinical science. *Anesthesiology.* 2020;132(1):55−68. Available from: https://doi.org/10.1097/ALN.0000000000002956.

12. Terrando N, Brzezinski M, Degos V, et al. Perioperative cognitive decline in the aging population. *Mayo Clin Proc*. 2011;86(9):885–893. Available from: https://doi.org/10.4065/mcp.2011.0332.
13. Woolf SH. The meaning of translational research and why it matters. *JAMA*. 2008;299(2):211–213. Available from: https://doi.org/10.1001/jama.2007.26.
14. Keramaris NC, Kanakaris NK, Tzioupis C, Kontakis G, Giannoudis PV. Translational research: from benchside to bedside. *Injury*. 2008;39(6):643–650. Available from: https://doi.org/10.1016/j.injury.2008.01.051. Epub May 27, 2008.
15. Hovens IB, Schoemaker RG, van der Zee EA, Heineman E, Izaks GJ, van Leeuwen BL. Thinking through postoperative cognitive dysfunction: how to bridge the gap between clinical and pre-clinical perspectives. *Brain Behav Immun*. 2012;26(7):1169–1179. Available from: https://doi.org/10.1016/j.bbi.2012.06.004. Epub Jun 21, 2012.
16. Cascella M, Muzio MR, Bimonte S, Cuomo A, Jakobsson JG. Postoperative delirium and postoperative cognitive dysfunction: updates in pathophysiology, potential translational approaches to clinical practice and further research perspectives. *Minerva Anestesiol*. 2018;84(2):246–260. Available from: https://doi.org/10.23736/S0375-9393.17.12146-2. Epub October 4, 2017.
17. Guéniot L, Lepere V, De Medeiros GF, et al. Muscle injury induces postoperative cognitive dysfunction. *Sci Rep*. 2020;10(1):2768. Available from: https://doi.org/10.1038/s41598-020-59639-3.
18. Borozdina A, Qeva E, Cinicola M, Bilotta F. Perioperative cognitive evaluation. *Curr Opin Anaesthesiol*. 2018;31(6):756–761. Available from: https://doi.org/10.1097/ACO.0000000000000658.
19. Bilotta F, Lauretta MP, Borozdina A, Mizikov VM, Rosa G. Postoperative delirium: risk factors, diagnosis and perioperative care. *Minerva Anestesiol*. 2013;79(9):1066–1076.
20. Fadayomi AB, Ibala R, Bilotta F, Westover MB, Akeju O. A systematic review and *meta*-analysis examining the impact of sleep disturbance on postoperative delirium. *Crit Care Med*. 2018;46(12):e1204–e1212. Available from: https://doi.org/10.1097/CCM.0000000000003400.
21. Bilotta F, Giordano G, Sergi PG, Pugliese F. Harmful effects of mechanical ventilation on neurocognitive functions. *Crit Care*. 2019;23:273. Available from: https://doi.org/10.1186/s13054-019-2546-y.
22. Chen C, Zhang Z, Chen T, Peng M, Xu X, Wang Y. Prolonged mechanical ventilation–induced neuroinflammation affects postoperative memory dysfunction in surgical mice. *Crit Care*. 2015;19(1):159. Available from: https://doi.org/10.1186/s13054-015-0882-0.
23. Turon M, Fernandez-Gonzalo S, de Haro C, et al. Mechanisms involved in brain dysfunction in mechanically ventilated critically ill patients: implications and therapeutics. *Ann Transl Med*. 2018;6(2):30. Available from: https://doi.org/10.21037/atm.2017.12.10.
24. Saczynski JS, Marcantonio ER, Quach L, et al. Cognitive trajectories after postoperative delirium. *N Engl J Med*. 2012;367(1):30–39.
25. Jackson WM, Gray DB, Jiang D, et al. Molecular mechanisms of anesthetic neurotoxicity: a review of the current literature. *J Neurosurg Anesthesiol*. 2016;28(4):361–372.
26. Xie Z, Dong Y, Maeda U, et al. The common inhalation anesthetic isoflurane induces apoptosis and increases amyloid beta protein levels. *Anesthesiology*. 2006;104(5):988–994. Available from: https://doi.org/10.1097/00000542-200605000-00015. PMID: 16645451.
27. Tanzi RE, Bertram L. Twenty years of the Alzheimer's disease amyloid hypothesis: a genetic perspective. *Cell*. 2005;120:545–555.
28. Le Freche H, Brouillette J, Fernandez-Gomez FJ, et al. Tau phosphorylation and sevoflurane anaesthesia: an association to postoperative cognitive impairment. *Anesthesiology*.

2012;116(4):779–787. Available from: https://doi.org/10.1097/ALN.0b013e31824be8c7. PMID: 22343471.

29. Bueé L, Bussiére T, Bué-Scherrer V, Delacourte A, Hof PR. Tau protein isoforms, phosphorylation and role in neurodegenerative disorders. *Brain Res Brain Res Rev*. 2000;33(1): 95–130.
30. Chander D, Garcia PS, MacColl JN, Illing S, Sleigh JW. Electroencephalographic variation during end maintenance and emergence from surgical anaesthesia. *PLoS One*. 2014;9:e106291 [PubMed:25264892].
31. Hofste WJ, Linssen CA, Boezeman EH, Hengeveld JS, Leusink JA, de-Boer A. Delirium and cognitive disorders after cardiac operations: relationship to pre- and intraoperative quantitative electroencephalogram. *Int J Clin Monit Comput*. 1997;14:29–36.
32. Bennett C, Voss LJ, Barnard JP, Sleigh JW. Practical use of the raw electroencephalogram waveform during general anaesthesia: the art and science. *Anaesth Analg*. 2009;109:539–550.
33. Fritz BA, Kalarickal PL, Maybrier HR, et al. Intraoperative electroencephalogram suppression predicts postoperative delirium. *Anesth Analg*. 2016;122(1):234–242. Available from: https://doi.org/10.1213/ANE.0000000000000989.
34. Ching S, Purdon PL, Vijayan S, Kopell NJ, Brown EN. A neurophysiological-metabolic model for burst suppression. *Proc Natl Acad Sci USA*. 2012;109:3095–3100.
35. Morris JC, Price JL. Pathologic correlates of non-demented aging, mild cognitive impairment, and early-stage Alzheimer's disease. *J Mol Neurosci*. 2001;17:101–118.
36. Brown CH, Jones EL, Lin C, et al. Shaping anaesthetic techniques to reduce postoperative delirium (SHARP) study: a protocol for a prospective pragmatic randomized controlled trial to evaluate spinal anaesthesia with targeted sedation compared with general anaesthesia in older adults undergoing lumbar spine fusion surgery. *BMC Anaesthesiol*. 2019;19:192. Available from: https://doi.org/10.1186/s12871-019-0867-7.
37. Patel V, Champaneria R, Dretzke J, Yeung J. Effect of regional vs general anaesthesia on postoperative delirium in elderly patients undergoing surgery for hip fracture: a systematic review. *BMJ Open*. 2018;8(12):e020757. Available from: https://doi.org/10.1136/bmjopen-2017-020757. PMID: 30518580; PMCID: PMC6286489.
38. Canet J, Raeder J, Rasmussen LS, et al. Cognitive dysfunction after minor surgery in the elderly. *Acta Anaesthesiol Scand*. 2003;47(10):1204–1210. Available from: https://doi.org/10.1046/j.1399-6576.2003.00238.x.
39. Ferraris VA, Ballert EQ, Mahan A. The relationship between intraoperative blood transfusion and postoperative systemic inflammatory response syndrome. *Am J Surg*. 2013; 205(4):457–465. Available from: https://doi.org/10.1016/j.amjsurg.2012.07.042. Epub 2013 Feb 12. PMID: 23414633.
40. Tsuruta R, Oda Y. A clinical perspective of sepsis-associated delirium. *J Intensive Care*. 2016;4:18. Available from: https://doi.org/10.1186/s40560-016-0145-4. Published March 23, 2016.
41. Davenport RD. Pathophysiology of hemolytic transfusion reactions. *Semin Hematol*. 2005;42(3):165–168. Available from: https://doi.org/10.1053/j.Seminhematol.2005.04.006. PMID: 16041666.
42. Urner M, Herrmann IK, Buddeberg F, et al. Effects of blood products on inflammatory response in endothelial cells in vitro. *PLoS One*. 2012;7:e33403.
43. van Gool WA, van de Beek D, Eikelenboom P. Systemic infection anddelirium: when cytokines and acetylcholine collide. *Lancet*. 2010;375:773–775.

44. Willard LB, Hauss-Wegrzyniak B, Wenk GL. Pathological and biochemical consequences of acute and chronic neuroinflammation within the basalforebrain cholinergic system of rats. *Neuroscience*. 1999;88:193–200.
45. Zhu SH, Ji MH, Gao DP, Li WY, Yang JJ. Association between perioperative blood transfusion and early postoperative cognitive dysfunction in aged patients following total hip replacement surgery. *Ups J Med Sci*. 2014;119(3):262–267. Available from: https://doi.org/10.3109/03009734.2013.873502.
46. Behrends M, DePalma G, Sands L, Leung J. Association between intraoperative blood transfusions and early postoperative delirium in older adults. *J Am Geriatr Soc*. 2013;61(3):365–370. Available from: https://doi.org/10.1111/jgs.12143. PMID: 23496244; PMCID: PMC4529289.
47. Inouye SK, van Dyck CH, Alessi CA, et al. Clarifying confusion: the Confusion Assessment Method. A new method for detection of delirium. *Ann Intern Med*. 1990;113:941–948.
48. Inouye SK. Prevention of delirium in hospitalized older patients: risk factors and targeted intervention strategies. *Ann Med*. 2000;32(4):257–263.
49. Marcantonio ER, Goldman L, Orav EJ, et al. The association of intraoperative factors with the development of postoperative delirium. *Am J Med*. 1998;105:380–384.
50. Whitlock EL, Vannucci A, Avidan MS. Postoperative delirium. *Minerva Anestesiol*. 2011;77:448–456.
51. Kehlet H. Enhanced postoperative recovery: good from afar, but far from good? *Anaesthesia*. 2020;75(Suppl. 1):e54–e61.
52. Belrose JC, Ruediger R. Anaesthesiology and cognitive impairment a narrative review of current clinical literature. *BMC Anesthesiol*. 2019;19:24. Available from: https://doi.org/10.1186/s12871-019-0903-7.
53. Denayer T, Stohr T, van Roy M, et al. Animal models in translational medicine: validation and prediction. *N Horiz Transl Med*. 2014;2(1):5–11.

Geriatrics and translational research

7

Suparna Bharadwaj and Sangeetha R. Palaniswamy
Department of Neuroanaesthesia and Neurocritical Care, National Institute of Mental Health and Neurosciences, Bengaluru, India

"The first requisite for immortality is death" said by Stanislaw Jerzy Lec, a Polish poet. His words will soon be contradicted as the advances in geriatric sciences contemplate directly halting the process of aging instead of disease-specific interventions.

Aging contributes primarily to morbidity, mortality, and huge health costs to any nation.[1]

Geriatric science or geroscience is a multidisciplinary subject that deals with the molecular mechanism of aging and age-related health issues.[2] Despite significant advances in gerosciences, 20th-century approaches have been to treat or prevent age-related diseases or disorders. Such attempts have been relatively incompetent to adequately enhance the quality and quantity of life. Recent findings propose it is reasonable to mitigate age-related diseases by intervening intrinsic aging process. Many such interventions have almost reached the point of final human clinical trials. Clinical interventions targeting fundamental mechanisms of aging may prolong, avert, or treat age-related diseases and disabilities as a group, instead of one at a time.[3] In this chapter we will discuss (1) cellular, biochemical, and molecular biology of aging; (2) the physiology, immunology, and endocrinology of aging; (3) mechanisms and intervention of cellular senescence and stem cell aging; (4) computational and functional genomics of aging; (5) animal models of aging research; (6) geroprotectors as antiaging chemicals; and (7) human experiments on longevity

Cellular, biochemical, and molecular biology of aging

To accomplish longevity in humans, understanding cellular programs regulating the process of aging and how their dysregulation leads to functional decline and senility is vital. Here we review some of these critical functions and their dysregulation. Cellular health is controlled by organization of nuclear chromosomal structure, regulation of transcription, and import/export of substances from the nucleus.[4] Other factors controlling cellular health are protein translation, autophagic recycling of organelles, and maintenance of cytoskeletal structure and extracellular matrix (ECM).[5] Ultimately there is an intricate interplay of regulation controlling the aging of the cell as the regulatory system receives inputs from different other systems.

Perioperative Neuroscience. DOI: https://doi.org/10.1016/B978-0-323-91003-3.00018-0

Chromosome and telomere regulation

Repetitive sequence of telomeres is normally capped to every chromosome. Such a telomere sequence protects chromosomal damage caused by shortening at each replicative cycle.[6] Shortening is caused by problems on lagging strand and oxidative damage. Telomere length is inversely correlated with life span.[7]

Transcription regulation

Transcriptional regulation is critical in the activation and production of many genes which contribute to life span extension. Many molecular mechanisms that regulate longevity are at transcriptional signaling of IIS and target of rapamycin (TOR) pathways.[8]

Nuclear trafficking and organization

Eukaryotic nuclear pore complex (NPC) is a molecular device required for exporting proteins into and out of the nucleus. Nuclear trafficking reduces with cellular aging. NPCs are susceptible to age-related changes. Gradual decline of NPCs further contributes to aging through leakage of proteins.[9] Nuclear organization is also important for cellular longevity. Lamins spatially organize the genomes inside the nucleus. The incorrect organization of lamins causes laminopathies resulting in premature aging and age-related disorders.[10]

Protein translation

Protein translation is an important control mechanism in longevity regulation.[11] Downregulation of translation extends the life span in many organisms.

Proteostasis

Proteostasis is the maintenance of intracellular protein quality. Proteostasis ensures the health and longevity of the cell by the supply of high-quality protein by replacing misfolded and damaged proteins from the cell with newly formed healthy proteins.[5] Several age-related disorders, like Parkinson's disease, Alzheimer's disease, and amyotrophic lateral sclerosis, are associated with aggregated and misfolded proteins.

Unfolded protein response

Unfolded protein response (UPR) monitors the quality of translation and unfolded amino acid chains in the endoplasmic reticulum.[5] UPR dysregulation is associated with age-related disorders, like Parkinson's disease and Alzheimer's disease.

Autophagy

Autophagy is similar to proteostasis where the organelles are consumed by the native cell. Autophagy is essential for longevity paradigms across many species. Autophagic clearing of damaged proteins, protein aggregates, and organelles is needed to supply new raw material for a healthy cell. Autophagy plays a complex role in neurodegenerative disease.[12]

Mitochondrial function, biogenesis, and mitophagy

The maintenance of mitochondrial homeostasis is critical for cellular health. Mitochondria play a vital role in the energy generation and production of free radicals that cause damage. Defects in energy-generation pathways of oxidative phosphorylation or neutralization of free radicals result in various age-related diseases and reduced life span.[13]

Cytoskeletal integrity

Cytoskeleton is vital in maintaining cell shape and integrity and its malfunction is an indicator of cellular aging. Cytoskeletal damage can result in degenerative neural diseases characteristic of aging.[14] The ECM is a critical contributor to health and longevity and is also an indicator of intracellular health.[5]

Replicative aging senescence and renewal

Cells experience aging linked to the number of times they have undergone division. Such a phenomenon is called replicative aging. Replicative aging can be intervened by disruption of TOR signaling, dietary restriction, and changes in intracellular pH.[15] Stem cells maintain a balance between multipotency and tumorigenicity. Cellular senescence is a phenomenon of cell cycle arrest. An accumulation of senescent cells in the tissue is the cause of aging.[16] Stem cell exhaustion is another risk factor for the process of aging.[17]

Gaps in our knowledge

There is no single unified molecular mechanism of aging that reflects a composite of different theories. Cross talk has been established between various homeostatic pathways.

Physiology, immunology, and endocrinology of aging

Physiological changes: changes in nervous system

Neuronal dendrites undergo regression in number, length, and branching, and thus brain loses its mass. These morphological changes impair nerve fiber conduction contributing to age relative cognitive dysfunction.[18] Age-related loss of myelin integrity disrupts neural network synchrony. Cerebrospinal fluid turnover decreases in older adults and affects the clearance of metabolic products and the supply of nutrients. Age-related disruption of the blood–brain barrier is also observed.[19] Positron emission tomographic imaging reveals an age-related decrease in cortical serotonin receptors. Dopamine concentration and dopamine receptors are decreased in aged individuals. Diminished cholinergic functions have been described with aging. This highlights the importance of avoiding ant cholinergic medications in the elderly. There is an age-related decrease in the expression of NMDA receptors as well as a decrease in downstream intracellular signaling and neuronal responses.[20] Older individuals have an increased sensitivity of benzodiazepines to GABA-A receptors. It has been demonstrated that increasing histamine in the aging brain exaggerates age-related cognitive deficits.

Changes in cardiovascular system

Age-related increase in collagen, decrease in elastin, glycosylation of proteins, and free radical damage cause vascular stiffness. Vascular stiffness causes myocardial diastolic dysfunction. Aging causes decreases in response to stimulation of β-receptors.[21] This attenuated response is due to changes in the second messenger function. Age-related decreased expression of α-1A and α-1D receptors responsible for contractile function has been noted. This may be an adaptation to cardiac hypertrophy. Age-related α2-receptor downregulation is responsible for the development of hypertension. Sympathetic tone dominates and vagal tone reduces with age. Vasopressin and brain natriuretic peptide levels increase with aging.

Changes in respiratory system

Lung function declines with age. Chest wall stiffens decreasing the static chest wall compliance.[22] Lung compliance increases because of loss in lung parenchymal elasticity with aging. Gas exchange decreases with aging.

Renal, metabolic, and endocrine aging

Age-related pharmacokinetic and pharmacodynamic changes occur for various drugs. Major alterations of pharmacokinetics in the aged are the result of the reduction in hepatic blood flow, liver size and not due to age-related changes in hepatic enzymes responsible for drug metabolism. There will also be an age-related decrease in renal blood flow, glomerular filtration rate, and creatinine clearance.[23] Insulin resistance, diabetes mellitus, thyroid abnormalities, and a decrease in sex hormones are age-related endocrine changes.

Age-related immunological changes

The aging of the immune system results in dramatic alterations in the distribution and proficiency of immune cells. With aging, there is a loss of adaptive immunity and a gain of nonspecific innate immunity.[24] As a consequence, older individuals are susceptible to infection, cancer, and chronic inflammation. There is also a deterioration of regulatory T-cell function.

Mechanisms and intervention of cellular senescence and stem cell aging

Cellular senescence is defined as a state of replicative arrest, suppressed apoptosis that occurs with aging. Cellular senescence is associated with a secretory profile senescence associated secretory phenotype.[16] One of the most common causes of cellular replicative arrest or senescence is telomere shortening. Telomere shortening produces DNA damage and in an attempt to repair such damage, cellular replication is arrested. Another cause of cellular senescence is the accumulation of unfolded/misfolded endoplasmic reticulum proteins. Accumulation of senescent cells shortens life span by causing age-related changes in vital organs. Stem cells and antiaging genes like Klotho delay the aging process.[25]

Loss of stem cells and deficiency of Klotho gene are responsible for aging. Gradual loss of stem cell function occurs secondary to age-related cellular accumulation of toxic substances. Age-related loss of functional stem cell pool has been observed with skeletal muscle stem cells, neural stem cells, and germline stem cells. Stem cell depletion may be due to an imbalance between stem cell quiescence and proliferation. Stem cells either undergo apoptosis or senescence induced by exposure to cellular stress. Age-related diminution of extracellular signals from niche cells can also lead to stem cell aging. Interventions to arrest or prevent stem cell aging include stem cell transplantation, reprograming of somatic cells into the desired type of stem cell.[26] Resetting of epigenetic memory can reverse aging in stem cells.

Computational and functional genomics of aging

Computational and functional genomics use computational tools to construct the circuitry of RNA regulation and its cross talk with other nuclear and extranuclear circuits. Computational modeling of coding and noncoding polymorphism and RNA dynamics across the aging process is also in the present-day research arena.[27] A wealth of publicly available data on aging is the backbone for computational models of geroscience research (https://genomics.senescence.info/links.html). Easy access to these data and their computational analysis is critical to understand the cause and effect of aging. Bioinformatics is vital from data handling to interpretation and integration to complex comparisons, hypotheses, and model building.

Animal models of aging research

Table 7.1 shows animal models that were used in phase 0 trials of aging research.[28]

Table 7.1 Animal models of ageing research.

Animal	Type of breed	Advantage	Disadvantage	Type of study
Inbred mice	Breeding between relatives C57BL/6 mice	Minimal genetic variability, outcome can be attributed to environmental or treatment effects	Outcome may not represent the species	Cancer, aging, cardiovascular disorders, obesity, infectious disease susceptibility, blood disorders, neurosensory disorders, drug addiction, and toxicity
Outbred and F1 mice	BALB/cJ x C57BL/6 JF1 mice	Hybrid vigor, with long life spans, high disease resistance, early fertility, large and frequent litters, rapid growth, and large size. Outbred mice are more representative of the human population	Cannot be used to determine sensitivities to substances or examining physiological parameters	Caloric and methionine restriction as an intervention to extend the life span
Wild-derived mice	–			Calorie restriction extends life span
Genetically heterogeneous mouse models	HET3 mice	Each mouse is genetically unique	–	Calorie restriction extends life span in HET 3 mice
Knockout mice		Multiple genes can be knocked out	–	Various age-related disease models
Delayed aging mice models	Ames dwarf, Snell dwarf, and growth hormone receptor knockout mice	Lower core body temperature, improved insulin sensitivity, enhanced antioxidant defenses, and delayed onset of neoplasia	–	Klotho gene alteration

Transgenic rat model	Alzheimer's disease model, P23H transgenic albino rat	Disease model is closer to humans	–	Alzheimer's disease, retinitis pigmentosa
Life span extension models	GH knockout rats	–	–	Life span expansion
Naked mole rats	NMR	highly socialized	–	Behavioral, neurological, and physiological research
Primates	Rhesus monkey	–	Intelligent and aggressive animals, difficult to maintain	
Fish	Zebra fish, kill fish	regenerative capabilities	–	Tissue repair and longevity
Dogs and cats	–	Aging phenotypes of dogs and cats are attractive models for age-related disorders		Dogs—cognitive deterioration, age-associated neurodegenerative Disorders Cats—kidney disease, arthritis, sarcopenia, and neurological decline
Birds	–	–	Inherent longevity—life span studies are difficult to perform	Comparative cross-species studies of the interplay of metabolism and aging

Geroprotectors as antiaging chemicals

Geroprotectors are drugs or pharmacological agents proposed to extend life span by decreasing the rate of aging.[29] More than 400 compounds have been used to test life span extension as per DrugAge database.[30] Aging is not considered a disease and hence very few clinical trials exist on geroprotectors.

Primary criteria for potential geroprotectors

Following are the primary criteria for potential geroprotectors:[29]

1. Drugs should prolong the life of the animal model beyond its natural predicted life span.
2. Slowing of progression of age-related changes should be evident in animal models or human subjects.
3. The dose of geroprotector should be less than the toxic dose.
4. There should be minimal side effects.
5. Potential geroprotector should produce initial improvement in physical, mental, and emotional health supporting the long-term use of such agents.

Secondary criteria for potential geroprotectors

Following are the secondary criteria for potential geroprotectors:[29]

1. Geroprotective effects should be reproducible.
2. Geroprotectors should delay the progress of age-related diseases in humans.
3. Geroprotectors should increase organism resistance to unfavorable environmental factors.

Classification of geroprotectors

Geroprotectors are classified based on the concept of homeostasis and homeodynamics.[29]

Group I: Suppress the consequences of hemodynamic disruption—antidiabetic, antiarrhythmic, lipid lowering, cardiovascular and antihypertensive drugs.

Group II: Ensures homeo-dynamics—caloric restriction, immunomodulators

Group III: Suppresses excessive homeo-dynamics reactions

Geroprotectors may also be classified as those which extend life span only or health span only or both health and life span. A mere extension of health span by a drug in mice without effects on life span will fail to extend life span in humans as well.

Various drugs have been studied in animals and their application in humans as potential geroprotectors.[31]

1. *Antioxidants*: Mortality is increased with antioxidants in humans, they promote cancer in mice.
2. *Calorie restriction (CR)*: CR delays age-related diseases in humans and hence extends life span. CR also extends the life span in mice, rats, and monkeys. But some gerontologists think that CR will not increase the life span in humans.
3. *Resveratrol*: Resveratrol did not extend life in three mice studies. But in one of them, it improved health without life span extension. It also delayed vascular aging in Wistar

rats. In a study from 2006 resveratrol increased life span in mice on a high-calorie diet but did not extend life span in mice on a standard diet.

4. *Rapamycin*: Many studies have shown that rapamycin extends life span in all strains of normal mice as well as in cancer-prone and short-lived mice. It showed life span extension consistently in all age groups. Rapamycin more than doubled the life span in short-lived mice. Extension of life span is directly proportional to the dose of rapamycin in mice. However, rapamycin shortened the life span of artificial mice lacking telomerase. Mechanistic TOR (mTOR) is a kinase in humans that is encoded by mTOR gene. It regulates cell growth, cell proliferation, cell motility, cell survival, protein synthesis, autophagy, and transcription. mTOR has been shown to drive several aspects of cerebrovascular dysfunction. Vascular cognitive impairment and dementia, blood–brain barrier breakdown, and impaired neurovascular coupling have been shown to be driven by mTOR. Chronic mTOR attenuation by rapamycin is known to preserve cerebrovascular function and microvascular integrity.
5. *Curcumin*: Curcumin did not show an extended life span in mice of 4 months and 12 months of age respectively.
6. *Quercetin*: Dietary supplement of quercetin did not extend life span in Fl hybrid mice. However, a senolytic combination of quercetin and dasatinib extended the median life span by 6.3%. In this study, C57BL/6 mice aged 24–27 months were given the senolytic combo every 2 weeks by oral gavage for 3 consecutive days.
7. *Spermidine*: Spermidine administered to C57BL/6 J mice extended medium life span by 10%.
8. *NAD boosters*: Nicotinamide riboside (NR) and nicotinamide (NAM) are NAD + boosters. NR extends life span in mice by 4.7%, but NAM did not extend life span in male C57BL/6 J mice. Short-lived mice with progeroid syndromes had dramatic increase in life span with NAD boosters. In normal mice, NAD boosters were not useful.
9. *Berberine*: mTOR pathway in cell culture is inhibited by berberine. However, it is not translated into the animal life span.
10. *Fisetin*: Fisetin extended life span in old mice but the sample size of this study was small.
11. *Metformin*: Life span extension effects are seen in female 129/Sv mice (by 4.4%), female SHR mice (mean life span by 37.8% & maximum life span by 10.3%). Metformin decreased life span in inbred 129/Sv male mice (by 13.4%). Combination of metformin and SRT1720, a sirtuin 1-activator decreased life span. In neonatal 129/Sv mice metformin extended life span in male but not in female mice. Low dose (0.1% w/w in diet) of metformin extended life span in C57BL/6 mice while high dose (1% w/w in diet) decreased life span. In genetically heterogenous mice a low dose of metformin alone did not extend life span. However, the combination of metformin and rapamycin (14 ppm) robustly extended the life span. Effects of metformin on life span are inconsistent in normal mice whereas it consistently increases life span in short-lived cancer-prone mice.
12. *Acarbose*: Acarbose, alpha-glucosidase inhibitor increased median longevity in male mice. A low carbohydrate diet in combination with rapamycin also increased life span.
13. *Angiotensin-converting enzyme in inhibitors*: Enalapril increased survival in mice. Losartan and enalapril extended life span in Wistar rats. Mean survival increased by 19% and 21% for losartan and enalapril respectively. Ramipril and simvastatin combination increased the life span of long-lived B6C3F1 male mice.
14. *Mitophagy*: Cellular house-keeping process that eliminates damaged mitochondria is called mitophagy. Drugs/measures to promote and sustain mitophagy may be therapy for age-related conditions, mainly cancer.
15. Other geroprotectors, their mechanism of action, and the current ongoing research are listed in Table 7.2.

Table 7.2 Other geroprotectors.

Geroprotector	Action	Disease against which used	Ongoing research
Deuterium reinforced linoleic acid	lowers lipid peroxidation and mitigates cognitive impairment	mouse model of Huntington's disease	
	Anti-amyloid interventions	No cognitive improvement	
Urilithin A—polyphenolic-rich plant food	Improves muscle health in old animals		Increases life span in *Caenorhabditis elegans* and in old mice In elderly modulates acylcarnitines in plasma and expression of skeletal muscle mitochondria genes-can be helpful to treat sarcopenia
Dasatinib and quercetin combination	Senolytic activity	Idiopathic pulmonary fibrosis	Pulmonary function and fragility index of study subjects did not change. There was significant improvement with endurance while walking, speed of movement and keeping the body in an upright position
UBX0101	Senolytic	Senolytic in mice, osteoarthritis	Mixed conclusions with human studies
TORC1 inhibitors	Regulation of cellular aging	Suppression of TORC 1 increases life span	Decrease in rate of infection in elderly subjects with low-dose combination of a catalytic (BEZ235) and allosteric (RAD001) TORC 1 inhibitor. The effect is observed for a year after drug initiation

(*Continued*)

Table 7.2 (Continued)

Geroprotector	Action	Disease against which used	Ongoing research
			Upregulation of antiviral gene expression and improvement in response to influenza vaccination in subjects on TORC 1 inhibitors Clinical trial of RTB101 an oral, selective TIRC 1 inhibitor was stopped at phase 3 as it did not meet the primary endpoint
Sanchi extract —hormetin product	Induced the transcription of stress genes and increased the synthesis of stress proteins, especially HSP1A1 in normal human keratinocytes and dermal fibroblasts	Skin care in healthy subjects	

Human experiments on longevity

Clinical doses of metformin are tolerated well in humans. As per several retrospective and prospective data of numerous studies, metformin decreases all-cause mortality in patients with different diseases.[32] Metformin decreases the incidence of cancer and a combination of metformin/rapamycin is shown to have potential benefits in humans. In humans, surrogate trials are conducted based on the fact that when health span is increased, life span also increases. Several drugs that extend life span in animals are already in use by humans. However, their role as life-extending drugs has not been proven yet. Fisetin, vitamin B3 and its analogs, aspirin, rapamycin, metformin, dasatinib, and enalapril are some of them.[31]

Data-BETA project is initiated by Josh Mitteldorf studying the impact of dietary supplements, dietary changes, and exercise regimes on biological age.[33] Methylation clocks are being evaluated as antiaging programs. Methylation clocks are epigenetic clocks comprising a set of nucleotide sites whose DNA methylation levels determine the chronological age of humans and other vertebrates. Data-BETA project proposes to study methylation clocks in 5000 persons to measure the effect of a wide range of interventions. This study will also observe the impact of the combination of interventions on aging.

Nicotinamide adenine dinucleotide (NAD) upregulation is studied in acute kidney injury, Alzheimer's disease, chronic fatigue syndrome, dementia, hyperphosphatemia, hypertension, obesity, Parkinson's disease, photoaging of skin, psoriasis, skin cancers, type 1 diabetes mellitus, type 2 diabetes mellitus, and schizophrenia. Human trials show that NAD upregulation is a promising treatment strategy for health span extension.[34] Exercise programs tailored to old individuals are some of the proposed approaches for NAD upregulation.

Reducing chronic inflammation of psoriasis has been associated with decreased atherosclerotic lesions. These findings have demonstrated that biological therapy to treat psoriasis is as beneficial as statin therapy on coronary arteries.[35]

Wide range of preclinical evidence proposes that short-term fasting and fasting-mimicking diets (FMDs) can mitigate the effect of stressors and chemotherapy on healthy cells.[36] At the same time, they render the cancer cells more susceptible to chemotherapy and other therapies. In the DIRECT trial, 90%–100% tumor cell loss occurred in patients using FMD (odds ratio 4.109). FMD also significantly inhibits chemotherapy-induced DNA damage in T lymphocytes.

Gut microbes comprise bacteria, fungi, viruses, and protozoa. They are known to modulate age-related changes in innate immunity, sarcopenia, and cognitive function which are important components of frailty syndrome.[37] Gut microbial flora was studied in 422,417 individuals from the UK biobank.[38] The scientists found out that the higher levels of 11 bacteria were associated with 28 health and disease outcomes like chronic obstructive pulmonary diseases (COPD), atopy, high blood pressure, and high blood lipids.

Health and survival advantages have been associated with low incidence of mental and behavioral disorders, tobacco-related cancers, low occurrence of teenage parenthood, early marriage, and divorce.

Utility of epigenetic clock in aging and age-related diseases

Epigenetic clock provides information on how much an individual's biological age differs from chronological age.[39,40] The epigenetic clocks are a methylation-based predictor of age (HorvathAge, HannumAge), all-cause mortality (DNAm PhenoAge, DNAm GrimAge) and telomere length (DNAm Telomere Length). DunedinPoAm is a speedometer epigenetic clock that is a single time point predictor of biological aging. These epigenetic clocks are associated with biological and chronological aging as well as several age-related diseases and mortality. They are associated with the prevalence of type 2 diabetes mellitus, COPD, lung cancer, and cardiovascular diseases.

Conclusion

The impending biological process of aging leads to an increased risk of chronic disease, disability, and related socioeconomic consequences. Presently, there is an unprecedented demand for extended health span and life span. To meet this demand, geoscientists constantly study aging from model organisms to older adults. They work in a highly integrative approach to reach a new dimension of translational geroscience. Translational geroscience is the new paradigm of 21st-century medicine where the hallmarks of aging are targeted directly rather than age-related diseases. Preclinical studies and demographic analysis indicate that the process of arresting aging is feasible. Although challenges exist, arresting aging in humans is extensively studied with great potential for significant increases in longevity and health span.

References

1. Lunenfeld B, Stratton P. The clinical consequences of an ageing world and preventive strategies. *Best Pract Res Clin Obstet Gynaecol.* 2013;27(5):643–659.
2. Fabbri E, Zoli M, Gonzalez-Freire M, Salive ME, Studenski SA, Ferrucci L. Aging and multimorbidity: new tasks, priorities, and frontiers for integrated gerontological and clinical research. *J Am Med Dir Assoc.* 2015;16(8):640–647.
3. Figueira I, Fernandes A, Mladenovic Djordjevic A, et al. Interventions for age-related diseases: shifting the paradigm. *Mech Ageing Dev.* 2016;160:69–92.
4. Zwerger M, Ho CY, Lammerding J. Nuclear mechanics in disease. *Annu Rev Biomed Eng.* 2011;13:397–428.
5. DiLoreto R, Murphy CT. The cell biology of aging. *Mol Biol Cell.* 2015;26(25): 4524–4531.
6. O'Sullivan RJ, Karlseder J. Telomeres: protecting chromosomes against genome instability. *Nat Rev Mol Cell Biol.* 2010;11(3):171–181.
7. Whittemore K, Vera E, Martínez-Nevado E, Sanpera C, Blasco MA. Telomere shortening rate predicts species life span. *Proc Natl Acad Sci U S A.* 2019;116(30): 15122–15127.
8. Pan H, Finkel T. Key proteins and pathways that regulate life span. *J Biol Chem.* 2017;292(16):6452–6460.
9. D'Angelo MA, Raices M, Panowski SH, Hetzer MW. Age-dependent deterioration of nuclear pore complexes causes a loss of nuclear integrity in postmitotic cells. *Cell.* 2009;136(2):284–295.
10. Schreiber KH, Kennedy BK. When lamins go bad: nuclear structure and disease. *Cell.* 2013;152(6):1365–1375. Available from: https://doi.org/10.1016/j.cell.2013.02.015.
11. Anisimova AS, Alexandrov AI, Makarova NE, Gladyshev VN, Dmitriev SE. Protein synthesis and quality control in aging. *Aging (Albany NY).* 2018;10(12):4269–4288.
12. Kiriyama Y, Nochi H. The function of autophagy in neurodegenerative diseases. *Int J Mol Sci.* 2015;16(11):26797–26812. Published 2015 Nov 9.
13. Phaniendra A, Jestadi DB, Periyasamy L. Free radicals: properties, sources, targets, and their implication in various diseases. *Indian J Clin Biochem.* 2015;30(1):11–26.

14. Kounakis K, Tavernarakis N. The cytoskeleton as a modulator of aging and neurodegeneration. *Adv Exp Med Biol*. 2019;1178:227–245.
15. Antikainen H, Driscoll M, Haspel G, Dobrowolski R. TOR-mediated regulation of metabolism in aging. *Aging Cell*. 2017;16(6):1219–1233.
16. McHugh D, Gil J. Senescence and aging: causes, consequences, and therapeutic avenues. *J Cell Biol*. 2018;217(1):65–77.
17. Sameri S, Samadi P, Dehghan R, Salem E, Fayazi N, Amini R. Stem cell aging in lifespan and disease: a state-of-the-art review. *Curr Stem Cell Res Ther*. 2020;15(4):362–378.
18. Peters A. The effects of normal aging on myelin and nerve fibers: a review. *J Neurocytol*. 2002;31(8–9):581–593.
19. Erickson MA, Banks WA. Age-Associated changes in the immune system and blood–brain barrier functions. *Int J Mol Sci*. 2019;20(7):1632.
20. Cleeland C, Pipingas A, Scholey A, White D. Neurochemical changes in the aging brain: a systematic review. *Neurosci Biobehav Rev*. 2019;98:306–319.
21. Ferrara N, Komici K, Corbi G, et al. β-adrenergic receptor responsiveness in aging heart and clinical implications. *Front Physiol*. 2014;4:396.
22. Sharma G, Goodwin J. Effect of aging on respiratory system physiology and immunology. *Clin Interv Aging*. 2006;1(3):253–260.
23. Shah S, Lewis MC. Renal, metabolic, and endocrine aging. In: Reves J, Barnett S, McSwain J, Rooke G, eds. *Geriatric Anesthesiology*. Cham: Springer; 2018. Available from: https://doi.org/10.1007/978-3-319-66878-9_13.
24. Montecino-Rodriguez E, Berent-Maoz B, Dorshkind K. Causes, consequences, and reversal of immune system aging. *J Clin Invest*. 2013;123(3):958–965.
25. Bian A, Neyra JA, Zhan M, Hu MC. Klotho, stem cells, and aging. *Clin Interv Aging*. 2015;10:1233–1243. Published 2015 Aug 4.
26. Oh J, Lee YD, Wagers AJ. Stem cell aging: mechanisms, regulators and therapeutic opportunities. *Nat Med*. 2014;20(8):870–880.
27. Wieser D, Papatheodorou I, Ziehm M, Thornton JM. Computational biology for ageing. *Philos Trans R Soc Lond B Biol Sci*. 2011;366(1561):51–63. Available from: https://doi.org/10.1098/rstb.2010.0286.
28. Mitchell SJ, Scheibye-Knudsen M, Longo DL, de Cabo R. Animal models of aging research: implications for human aging and age-related diseases. *Annu Rev Anim Biosci*. 2015;3:283–303.
29. Moskalev A, Chernyagina E, Kudryavtseva A, Shaposhnikov M. Geroprotectors: a unified concept and screening approaches. *Aging Dis*. 2017;8(3):354–363. Published 2017 May 2.
30. Janssens GE, Houtkooper RH. Identification of longevity compounds with minimized probabilities of side effects. *Biogerontology*. 2020;21(6):709–719.
31. Trendelenburg AU, Scheuren AC, Potter P, Müller R, Bellantuono I. Geroprotectors: a role in the treatment of frailty. *Mech Aging Dev*. 2019;180:11–20.
32. Glossmann HH, Lutz OMD. Metformin and aging: a review. *Gerontology*. 2019;65(6):581–590.
33. Mitteldorf J. A clinical trial using methylation age to evaluate current antiaging practices. *Rejuvenation Res*. 2019;22(3):201–209.
34. Radenkovic D, Reason, Verdin E. Clinical evidence for targeting NAD therapeutically. *Pharm (Basel)*. 2020;13(9):247.
35. Caiazzo G, Fabbrocini G, Di Caprio R, et al. Psoriasis, cardiovascular events, and biologics: lights and shadows. *Front Immunol*. 2018;9:1668.

36. de Groot S, Lugtenberg RT, Cohen D, et al. Fasting mimicking diet as an adjunct to neoadjuvant chemotherapy for breast cancer in the multicentre randomized phase 2 DIRECT trial. *Nat Commun*. 2020;11(1):3083.
37. Nagpal R, Mainali R, Ahmadi S, et al. Gut microbiome and aging: physiological and mechanistic insights. *Nutr Healthy Aging*. 2018;4(4):267–285.
38. Groot HE, van de Vegte YJ, Verweij N, Lipsic E, Karper JC, van der Harst P. Human genetic determinants of the gut microbiome and their associations with health and disease: a phenome-wide association study. *Sci Rep*. 2020;10(1):14771.
39. Fransquet PD, Wrigglesworth J, Woods RL, Ernst ME, Ryan J. The epigenetic clock as a predictor of disease and mortality risk: a systematic review and *meta*-analysis. *Clin Epigenetics*. 2019;11(1):62.
40. Jiang S, Guo Y. Epigenetic clock: DNA methylation in aging. *Stem Cell Int*. 2020;2020:1047896.

Pediatrics and translational research

8

Kulikov Alexander[1], *Russi Mariarita*[2] *and Bilotta Federico*[2]
[1]Burdenko National Medical Research Center of Neurosurgery, Moscow, Russia,
[2]Department of Anesthesiology and Critical Care, Policlicnico Umberto I, "Sapienza" University of Rome, Rome, Italy

Introduction

Translational research in perioperative pediatric neuroscience is intended to incorporate into the clinical practice the available knowledge derived from laboratory researches related to the influence of surgical interventions and anesthetic techniques on neurocognitive human development from birth till adulthood.[1] The primary goal of this interdisciplinary branch of biomedical science is to optimize pediatric patient management in order to improve screening, prevention, and treatment with the ultimate aim of improving health care policies.[2] The main focus of the researches in this field is to prevent and assess the potential damage induced by anesthetics on the developing brain.[3] In this context, preclinical studies were designed to evaluate anesthetics' safety and possible underlying neurotoxic mechanisms.[4] Collected findings served as background rationale for clinical trials that lead to contrasting results.[5] Despite existing ambiguity, which makes it difficult to translate preclinical research findings to the community and routine clinical practice, the United States Food and Drug Administration (FDA) issued an official warning on "repeated or lengthy use of general anesthetic and sedation drugs during surgeries or procedures in children younger than 3 years or in pregnant women during their third trimester may affect the development of children's brains."[6] Therefore the role and validity of translational research in perioperative pediatrics neuroscience are still controversial.

This chapter is intended to present preclinical and clinical evidence related to translational research in perioperative pediatric neuroscience with the focus on perioperative anesthetics neurotoxicity and to discuss the inconsistencies between these two settings.

Preclinical research on perioperative anesthetics neurotoxicity

Preclinical studies on the possible neurotoxic effects on the developing brain are based on multiple experimental approaches including both in vitro and in vivo

Perioperative Neuroscience. DOI: https://doi.org/10.1016/B978-0-323-91003-3.00006-4

models.[7] Tissue samples, cell lines, and three-dimensional brain organoids derived from pluripotent stem cells represent in vitro models. Investigation of cellular mechanisms of neurocytes growth and synapses formation under the influence of anesthetics are the main goals of such studies.[8,9] However, using injured tissue or disconnected population of cells in artificial environments under nonhomeostatic conditions are the main disadvantages of in vitro models. Therefore evidence on anesthetic toxicity extracted in those models carries substantial bias in extending to living being context. Because of this, in vivo models are more common and provide more reliable data on the potential threat of anesthetics on a developing brain. Among the tested animal models are rodents, guinea pigs, pigs, sheep, nonhuman primates, and nonmammalian species (nematode, drosophila, zebrafish, etc.).[10] The most reliable evidence was received in rodent models and confirmed in nonhuman primates. Hereafter, key findings of these studies will be presented according to four main factors, which are important for the real clinical practice: age of exposure; anesthetic choice; dose and duration of exposure; and adjuvant medication for neuroprotection.

Period of maximal brain vulnerability for anesthetics exposure is hypothesized to neuronal growth spurt by the majority of researches. Both in humans and in animals, active neuronal maturation takes usually a short period of time from fetal to early childhood and depends on the particular brain region development. In rodent models, brain vulnerability peaks at postnatal day 7 (PD7). The high susceptibility of neurons for the toxic effect of different anesthetics in this period was confirmed in a number of studies.[11–13] For instance, ketamine demonstrated high proapoptotic caspase-3 enzymes activation effect in PD7 rats, but not in PD14 or PD21 rats.[11] Reduced neurogenesis and increased level of proapoptotic enzymes were also recorded in PD7 rats under sevoflurane exposure.[12] Propofol decreased dendritic spine density at PD5 and PD10 rats, but in PD15, PD20, and PD30 it increased the dendritic spine density.[13] Isoflurane provoked progenitor cells proliferation and persistent hippocampal deficit when administered at PD7 but not in PD60 rats.[14] Of note, further studies confirmed that the proapoptotic effect of isoflurane is distributed among all brain regions in PD7 rats, while selectively involves the dentate girus and olfactory bulb in PD21 and PD49 rats. These data confirm different time-window vulnerability among the brain regions and a protracted susceptibility to neurotoxic effect of anesthetics, even in adult animals, where brain neurogenesis persists.[15] High vulnerability of developing brain to common anesthetics (ketamine, propofol, and isoflurane) was further confirmed also in nonhuman primates models where susceptibility peaks at PD5–PD6.[16–18]

Comparative preclinical studies on relative neurotoxicity of different anesthetics are sparse and inconsistent. Neurotoxicity of inhaled halogenates was tested in various studies with conflicting evidence: 6 h exposure of desflurane, isoflurane, or sevoflurane in experimentally determined equipotent anesthetic concentrations (7.3%, 1.6%, and 3.2%, respectively) were tested on a PD6 mice model and resulted in equivalent proapoptotic activity and neuronal cell death.[19] In a similar study design (PD6 mice exposed for 6 h) ran by a different research group, desflurane 8% led to greater impairment of working memory and more extensive neuroapoptosis than isoflurane 2% and sevoflurane 3%.[20] When tested in PD7 mice, 6-h inhalation of isoflurane 0.75% caused significantly greater apoptosis in the hippocampus and cortex than sevoflurane 1.1%.[21] Similarly, data on comparative neurotoxicity of intravenous or inhaled anesthetics are controversial: propofol anesthesia in PD7 rats induces more severe neurodegeneration and persistent learning deficit compared to control and sevoflurane treated animals.[22] A more recent study compared the effects of propofol to those of sevoflurane in combination with fentanyl or dexmedetomidine in PD21 and adult rats under controlled physiological conditions and proved that neuronal cell death was limited and similar in the four study groups.[23]

Other aspects of the possible role of the anesthetic regimens in the severity of neurotoxicity—including duration of exposure, the dose of anesthetics, and the number of exposures—have been addressed in preclinical studies. For instance, one of the earliest studies on anesthetic neurotoxicity tested the effects of extreme time exposure of pregnant rats to the halothane-enriched atmosphere (10 ppm for 8 h/day and 5 days/week throughout the pregnancy).[24] Not surprisingly, recorded results proved extensive and persisting neurodegeneration in the neonatal cortex of newborn rats. A subsequent seminal study, published in 2003, neuroapoptosis and persisting learning and memory disabilities in PD7 rats were induced by a combination of midazolam, nitrous oxide, and isoflurane administrated in doses sufficient to maintain a clinical equivalent depth of anesthesia for 6 h, reproducing a model that is closer to clinical practice.[25] Severity of neurotoxicity depending on the different duration of exposure was assessed in a number of studies. For instance, in PD16 rats inhaled anesthetics -isoflurane (1.5%), sevoflurane (2.5%), or desflurane (7%) increased dendritic spine density only after 6 h of exposure, but up to 2 h inhalation did not cause any signs of neuroapoptosis.[26] Ketamine anesthesia shorter than 3 h seemed to have few adverse effects, whereas exposures lasting from 9 to 24 h caused a significant increase in neuronal cell death in PD5–6 monkeys.[27] In contrast to the duration of exposure, the dose of anesthetics does not strictly correlate to the severity of neurotoxicity. For example, in adult mice, lower doses of isoflurane provoked more neurodegeneration and learning impairment compared to higher isoflurane concentrations (1% vs 1.5 or 2% for 1-h inhalation).[28] While the debate on the relevance of the dose and duration of tested anesthetic regimens is still open, it is now clear that multiple anesthesia exposures are proven risk factors of severe neurotoxicity in animal models. In a comparative study, four anesthetic regimens were assessed on rats model: (1) a single 2 h exposure to sevoflurane 2.5% on PD7; (2) a single 6 h exposure to sevoflurane 2.5% on PD7; (3) a repeated 2 h exposure to sevoflurane 2.5% for a total of 6 h on PD7, PD10, and PD13; (4) repeated 2 h exposure to 30% oxygen on the same days.[29] Recorded results confirmed that both single and multiple sevoflurane exposures can cause neurodegeneration and lead to long-lasting and dose-related ultrastructural damage, but the risk is especially high among rats with multiple exposures.

Neuroprotective adjuvants administration to mitigate the negative impact of anesthesia on the developing brain is a rather new issue for preclinical studies in the field. Among tested agents are xenon, dexmedetomidine, and insulin. Xenon

70% has a proapoptotic effect but alleviates the neuroapoptotic effect of isoflurane 0.75% in PD7 mice.[30] Dexmedetomidine, administered intraperitoneally at 25, 50, or 75 μg/kg in PD7 rats subsequently also exposed to isoflurane 0.75%, decreases neurodegeneration caused by the halogenate.[31] Of note, increased doses of dexmedetomidine to 50 and 75 μg/kg were not associated with more protection. Preoperative intranasal administration of insulin 7.0 μL before 3 h of sevoflurane 3%–5% inhalation was tested on PD7 mice model and prevented from long-term anesthesia-induced cognitive impairment.[32]

In summary, the evidence presented above proves the brain vulnerability of neonatal rodents and monkeys for anesthetic exposure, which can cause neuroapoptosis as well as persistent behavior disturbances. Duration of exposure and multiple exposures increase the severity of neurotoxicity. Data on relative neurotoxicity of different anesthetics and potential neuroprotection are limited which prevents clear conclusions based on preclinical data.

Clinical research on perioperative anesthetics neurotoxicity in children

Clinical interpretation of anesthetic neurotoxicity has been changed over time. In a narrative review on the neurotoxicity of anesthetics published in 1979 authors discussed this problem mainly in the context of brain irritation due to overdosing.[33] New insight in the clinical relevance of this issue emerged from the preclinical data described in the preclinical setting. Based on these findings, clinical studies were designed to evaluate the existence of long-term cognitive decline in pediatric patients exposed to anesthesia in early childhood. Initial studies were retrospective and compared the surrogate cognitive parameters (mainly, academic performance and school achievements) of children exposed and not exposed for anesthesia from big datasets: twins cohorts, regional birth children registers, or siblings cohorts. Only in recent years, the evidence based on prospective studies that used advanced psychological assessment has been accumulated. Hereafter some of these findings are presented.

A twin cohort analyzed through the Netherlands Twin Registry, containing 1143 pairs of monozygotic twins, is one of the first reliable attempts for assessing the impact of anesthesia on cognitive development in children.[34] Based on the parent's survey, children from this dataset were divided into three groups: twins who were "concordant for exposure" (both were exposed) to anesthesia in early childhood, twins who were "concordant for no anesthesia exposure" (none were exposed) and twins who were "discordant for anesthesia exposure" (only one of the twins was exposed to anesthesia). Primary outcome measures included standardized school tests and teacher ratings at the age of 12 years. Despite the fact that children exposed to anesthesia before 3 years had significantly lower educational achievement scores, there was no difference in cognitive assessment between twins discordant for anesthesia exposure. Authors concluded that there is no evidence for a causal relationship between anesthesia administration in early childhood and learning-related outcomes.

The analysis of regional birth children registers produced more powerful but less selective retrospective studies.[35,36] A group of 2689 children exposed to anesthesia for inguinal hernia repair before 1 year of age was compared to the age-matched control group. The study proved no differences in academic performance between anesthetized and control children, but the authors underlined that children exposed to anesthesia were more frequently unable to sit for the test due to significant neurodevelopmental or behavioral problems.[35] In a cohort of 33,514 children with at least 1 anesthesia episode before 4 years of age was compared to a matched unexposed cohort. The difference in achieved school grades was statistically significant but with no clear impact: one anesthetic episode associated with a 0.41% lower score and three or more exposures to anesthesia associated with a 1.82% decrease in school scores. Authors also underlined that this difference was seriously less than the differences associated with sex or maternal educational level.[36]

Retrospective analysis of siblings cohorts led to similarly ambiguous conclusions.[37,38] Analysis of birth cohort from the New York State Medicaid dataset (1999 to 2005) revealed 304 children who underwent surgery when they were younger than 3 years. Comparison of this group with matched control indicated increasing the risk of subsequent diagnosis of developmental or behavior disorders up to 60% than in a similar group of siblings who did not undergo surgery.[37] In contrast, no difference in readiness for school was found between exposed and unexposed to anesthesia children in the study enrolled 2346 sibling pairs with only one child exposed for surgery till 5 years of age.[38]

Considering the open controversies left by retrospective studies, prospective dedicated trials were realized: Pediatric Anesthesia NeuroDevelopment Assessment) (PANDA), Mayo Anesthesia Safety in Kids (MASK), and General versus Spinal Anesthesia in children (GAS) are the most notable among such projects.[39–41] In the multicenter PANDA study, 105 sibling pairs with ASA I-II scores were enrolled between 2009 and 2015.[39] In each pair, one of the siblings had one episode of anesthesia before age of 3 years while the other was not exposed to anesthesia before that age. Global cognitive function (IQ), as well as behavioral outcome, were prospectively assessed at age of 8–15 years in all participants. The median duration of inhalational anesthesia in the exposed children group was 80 min. The mean IQ was similar in both groups and there was also no difference in IQ subdomains or behavior status between the groups. In the MASK study, 997 children born and living in the same place were enrolled.[40] Among these kids, before age of 3 years, 411 were never exposed to anesthesia, 380 underwent anesthesia once and 206 were exposed multiple times to anesthesia. Neuropsychological testing [full-scale intelligence quotient (FSIQ) score of the Wechsler Abbreviated Scale of Intelligence] was performed at adolescence. The median cumulative duration of anesthesia (sevoflurane in most cases) was 45 and 187 min in singly and multiply exposed children, respectively. There was no significant difference in global intelligence score between the three groups, but processing speed and fine motor abilities were worse in those that were exposed to multiple anesthesias when kids. In the multicenter GAS trial, 722 children were enrolled between 2007 and 2013.[41] Children were randomly allocated to receive awake-regional (spinal) or general (sevoflurane) anesthesia for inguinal hernia repair. The primary outcome measure was FSIQ on the Wechsler Preschool and Primary Scale of Intelligence at 5 years of age. An important feature of this RCT is the early age of participants, considering that children were enrolled when had <60 weeks' postmenstrual age. The median duration of general anesthesia in the sevoflurane group was 54 min. There was no difference in scores based on FSIQ between the two groups and the authors concluded that 1 h of exposure to general anesthesia during early infancy does not cause measurable neurocognitive or behavioral deficits at the age of 5 years.

In summary, clinical and epidemiological data, recorded in retrospective studies and confirmed in prospective trials, do not confirm an impact of anesthesia on long-term behavioral and cognitive abilities when children are exposed to inhalational anesthesia of about 1-h duration. Evidence on more prolonged and multiple exposures is sparse.

Reasons for discrepancy between preclinical and clinical data

Evidence from preclinical and clinical studies on potential anesthetic neurotoxicity in children is inconsistent. Most of the preclinical studies report significant neuroanatomical and behavioral deterioration in animals exposed to anesthetics in the developing age. In contrast, human studies do not support preclinical evidence and there is not conclusive evidence on the risk balance related to postponing surgery. The reasons for this discrepancy are widely discussed in the literature and here are presented some of the possible explanations on the limited translation of preclinical evidence into clinical practice. Among the most relevant aspects that might contribute to determining a limited translation of preclinical evidence, there are the duration of the vulnerable period, the complexity of the brain maturation process, heterogeneity in exposure schemes, and control of physiological parameters during anesthesia.[4]

The vulnerable period, as we mentioned above, typically refers to the neuronal growth spurt in the developing brain, but the speed and time course of neuronal maturation varies greatly among the species. Certainly, growth spurt and potential for neuroplasticity during human development have a much longer duration than in rats, whose lifespan is only 2 years.[7] In this regard, the direct shift of the neurodegeneration observed in 7-day-old rat brain on the 3 years of human child growth seems not fully valid. The interregional gap in the maturation of different brain structures and complexity of this process, which includes various events, such as axon and dendrites growth and synaptogenesis, are additional complicating variables in clinical translation. Of note, the complexity of the maturation process was deeply investigated in the last decades in laboratory research and can be matched at *http://www.translatingtime.net*, which is a dataset containing the corresponding data on neurodevelopment steps across mammalian species. Brain heterogenic maturation also causes a difference in neuropsychological and behavioral consequences of anesthesia exposure in a particular time and animal species and makes it difficult to define the most appropriate testing tools to assess the cognitive function and to extrapolate results of the animal tests on human development.

The difference in tested exposure schemes is another aspect that affects the translation of preclinical data. Prolonged on 4–6 h anesthesia is used in most animal studies, sometimes with several anesthetics and multiple exposures without prolonged pauses. At the same time in clinical studies, typical interventions in children last no longer than 1 h under monoanesthesia of inhaled anesthetic. This discrepancy might reflect a structural limitation of preclinical studies that sometimes are based on a methodological approach that drives the tested conditions beyond the clinical setting. This bias can also be partially explained by the necessity of researchers involved in preclinical studies to produce "positive evidence" that has a higher potential of being published. This process can have partially misled researchers and now there is the possible overestimate of the impact on the clinical practice of a potential threat.[5]

The last but not least reason for the difficult translation of preclinical findings into the clinical setting is the poor control of physiological variables during animal exposure to anesthesia, such as heart rate, blood pressure, level of consciousness, blood oxygen, and carbon dioxide level. Therefore data recorded in such studies can be criticized as it is not possible to distinguish the extent of neuronal injury caused by the anesthetic exposure and that induced by abnormal physiological conditions (cardiovascular, respiratory, temperature, feeding, and volume status, etc). An additional limitation of some preclinical studies is the absence of nociceptive stimuli, which can affect the neurodevelopment itself.

Presented barriers for relative translation of preclinical evidence to clinical practice should stimulate a standardization of study protocols and reporting of preclinical findings. For instance, the SmartTots consortium is promoting consensus features of future studies designing.[10] The expert panel insists on the implementation of reporting standards in laboratory studies, which should clearly describe the rationale for anesthesia exposure, present physiological data, experimental sample size, gender of the animals, nutritional status, etc. Such new approaches could make the research in the field of anesthetic neurotoxicity in the pediatric population more fruitful and closely connected with real clinical practice.

Conclusion

Translational researches in perioperative neuroscience in children mainly focus on anesthetic neurotoxicity. Despite some preclinical evidence have suggested a possible impact of anesthetics on neurodevelopmental abilities, clinical evidence does not support these concerns. Effective translation of preclinical findings in clinical practice in this field is very complicated mainly due to significant inconsistency of animal and human research models. Laboratory, usually rodent-based, studies suggested that anesthetics can be neurotoxic, especially during the early period of brain development. At the same time, human-based retrospective and prospective studies do not confirm the significant impact of short-term anesthetic exposure even in early childhood. Future studies, both clinical and laboratory, should be carefully designed to overcome the mismatches and produce reliable data on the clinical significance of anesthetic neurotoxicity in children, since general anesthetics anyway are necessary and will be needed in the future for patients, including young children, who need surgery or other painful and stressful procedures.

References

1. Garcia-Rill E. *Translational Neuroscience a Guide to a Successful Program.* Wiley-Blackwell; 2012.
2. Randall JC, Martin T, Ghahramani P, Bidaut L, Higgins PJ, Shahzad A. Translational medicine definition by the European Society for Translational Medicine. *N Horiz Transl Med.* 2014;2(3):86–88.
3. Kharasch ED. The challenges of translation. *Anesthesiology.* 2018;128(4):693–696.
4. Disma N, Mondardini MC, Terrando N, Absalom AR, Bilotta F. A systematic review of methodology applied during preclinical anesthetic neurotoxicity studies: important issues and lessons relevant to the design of future clinical research. *Paediatr Anaesth.* 2016;26 (1):6–36.
5. Vutskits L, Culley DJ. GAS, PANDA, and MASK: no evidence of clinical anesthetic neurotoxicity!. *Anesthesiology.* 2019;131(4):762–764.
6. FDA. FDA Drug Safety Communication: FDA review results in new warnings about using general anesthetics and sedation drugs in young children and pregnant women. <https://www.fda.gov/drugs/drug-safety-and-availability/fda-drug-safety-communication-fda-review-results-new-warnings-about-using-general-anesthetics-and>; 2016.
7. McCann ME, Soriano SG. Does general anesthesia affect neurodevelopment in infants and children? *BMJ.* 2019;367:1645–1649.
8. Vutskits L, Gascon E, Tassonyi E, Kiss JZ. Clinically relevant concentrations of propofol but not midazolam alter in vitro dendritic development of isolated gamma-aminobutyric acid-positive interneurons. *Anesthesiology.* 2005;102(5):970–976.
9. Mintz CD, Barrett KM, Smith SC, Benson DL, Harrison NL. Anesthetics interfere with axon guidance in developing mouse neocortical neurons in vitro via a γ-aminobutyric acid type A receptor mechanism. *Anesthesiology.* 2013;118(4):825–833.
10. Chinn GA, Pearn ML, Vutskits L, et al. Standards for preclinical research and publications in developmental anaesthetic neurotoxicity: expert opinion statement from the SmartTots preclinical working group. *Br J Anaesth.* 2020;124(5):585–593.
11. Gutierrez S, Carnes A, Finucane B, et al. Is age-dependent, ketamine-induced apoptosis in the rat somatosensory cortex influenced by temperature? *Neuroscience.* 2010;168 (1):253–262.
12. Fang F, Xue Z, Cang J. Sevoflurane exposure in 7-day-old rats affects neurogenesis, neurodegeneration and neurocognitive function. *Neurosci Bull.* 2012;28 (5):499–508.

13. Briner A, Nikonenko I, De Roo M, Dayer A, Muller D, Vutskits L. Developmental Stage-dependent persistent impact of propofol anesthesia on dendritic spines in the rat medial prefrontal cortex. *Anesthesiology*. 2011;115(2):282–293.
14. Stratmann G, Sall JW, May LD, et al. Isoflurane differentially affects neurogenesis and long-term neurocognitive function in 60-day-old and 7-day-old rats. *Anesthesiology*. 2009;110(4):834–848.
15. Deng M, Hofacer RD, Jiang C, et al. Brain regional vulnerability to anaesthesia-induced neuroapoptosis shifts with age at exposure and extends into adulthood for some regions. *Br J Anaesth*. 2014;113(3):443–451.
16. Creeley C, Dikranian K, Dissen G, Martin L, Olney J, Brambrink A. Propofol-induced apoptosis of neurones and oligodendrocytes in fetal and neonatal rhesus macaque brain. *Br J Anaesth*. 2013;110(Suppl 1):29–38.
17. Paule MG, Li M, Allen RR, et al. Ketamine anesthesia during the first week of life can cause long-lasting cognitive deficits in rhesus monkeys. *Neurotoxicol Teratol*. 2011;33 (2):220–230.
18. Coleman K, Robertson ND, Dissen GA, et al. Isoflurane anesthesia has long-term consequences on motor and behavioral development in infant rhesus macaques. *Anesthesiology*. 2017;126(1):74–84.
19. Istaphanous GK, Howard J, Nan X, et al. Comparison of the neuroapoptotic properties of equipotent anesthetic concentrations of desflurane, isoflurane, or sevoflurane in neonatal mice. *Anesthesiology*. 2011;114(3):578–587.
20. Kodama M, Satoh Y, Otsubo Y, et al. Neonatal desflurane exposure induces more robust neuroapoptosis than do isoflurane and sevoflurane and impairs working memory. *Anesthesiology*. 2011;115(5):979–991.
21. Liang G, Ward C, Peng J, Zhao Y, Huang B, Wei H. Isoflurane causes greater neurodegeneration than an equivalent exposure of sevoflurane in the developing brain of neonatal mice. *Anesthesiology*. 2010;112(6):1325–1334.
22. Bercker S, Bert B, Bittigau P, et al. Neurodegeneration in newborn rats following propofol and sevoflurane anesthesia. *Neurotox Res*. 2009;16(2):140–147.
23. Ibrahim RM, Krammer CW, Hansen TG, Kristensen BW, Vutskits L, Sørensen JA. Systemic physiology and neuroapoptotic profiles in young and adult rats exposed to surgery: A randomized controlled study comprising four different anaesthetic techniques. *Int J Dev Neurosci*. 2015;45:11–18.
24. Chang LW, Dudley Jr AW, Katz J. Pathological changes in the nervous system following in utero exposure to halothane. *Env Res*. 1976;11(1):40–45.
25. Jevtovic-Todorovic V, Hartman RE, Izumi Y, et al. Early exposure to common anesthetic agents causes widespread neurodegeneration in the developing rat brain and persistent learning deficits. *J Neurosci*. 2003;23(3):876–882.
26. Briner A, De Roo M, Dayer A, Muller D, Habre W, Vutskits L. Volatile anesthetics rapidly increase dendritic spine density in the rat medial prefrontal cortex during synaptogenesis. *Anesthesiology*. 2010;112(3):546–556.
27. Zou X, Patterson TA, Divine RL, et al. Prolonged exposure to ketamine increases neurodegeneration in the developing monkey brain. *Int. J Dev. Neurosci*. 2009;27(7):727–731.
28. Valentim AM, Di Giminiani P, Ribeiro PO, Rodrigues P, Olsson IA, Antunes LM. Lower isoflurane concentration affects spatial learning and neurodegeneration in adult mice compared with higher concentrations. *Anesthesiology*. 2010;113(5):1099–1108.
29. Amrock LG, Starner ML, Murphy KL, Baxter MG. Long-term effects of single or multiple neonatal sevoflurane exposures on rat hippocampal ultrastructure. *Anesthesiology*. 2015;22(1):87–95.

30. Cattano D, Williamson P, Fukui K, et al. Potential of xenon to induce or to protect against neuroapoptosis in the developing mouse brain. *Can J Anaesth*. 2008;55(7):429–436.
31. Sanders RD, Sun P, Patel S, Li M, Maze M, Ma D. Dexmedetomidine provides cortical neuroprotection: impact on anaesthetic-induced neuroapoptosis in the rat developing brain. *Acta Anaesthesiol. Scand*. 2010;54(6):710–716.
32. Dai CL, Li H, Hu X, et al. Neonatal exposure to anesthesia leads to cognitive deficits in old age: prevention with intranasal administration of insulin in mice. *Neurotox Res*. 2020;38(2):299–311.
33. Steen PA, Michenfelder JD. Neurotoxicity of anesthetics. *Anesthesiology*. 1979;50:437–453.
34. Bartels M, Althoff RR, Boomsma DI. Anesthesia and cognitive performance in children: no evidence for a causal relationship. *Twin Res. Hum. Genet*. 2009;12(3):246–253.
35. Hansen TG, Pedersen JK, Henneberg SW, et al. Academic performance in adolescence after inguinal hernia repair in infancy: a nationwide cohort study. *Anesthesiology*. 2011;114(5):1076–1085.
36 Glatz P, Sandin RH, Pedersen NL, Bonamy AK, Eriksson LI, Granath F. Association of anesthesia and surgery during childhood with long-term academic performance. *JAMA Pediatr*. 2017;171(1):1634–1670.
37. DiMaggio C, Sun LS, Li G. Early childhood exposure to anesthesia and risk of developmental and behavioral disorders in a sibling birth cohort. *Anesth. Analg*. 2011;113(5):1143–1151.
38. O'Leary JD, Janus M, Duku E, et al. Influence of surgical procedures and general anesthesia on child development before primary school entry among matched sibling pairs. *JAMA Pediatr*. 2019;173(1):29–36.
39. Sun LS, Li G, Miller TL, et al. Association between a single general anesthesia exposure before age 36 months and neurocognitive outcomes in later childhood. *JAMA*. 2016;315(21):2312–2320.
40. Warner DO, Zaccariello MJ, Katusic SK, et al. Neuropsychological and behavioral outcomes after exposure of young children to procedures requiring general anesthesia: the mayo anesthesia safety in kids (mask) study. *Anesthesiology*. 2018;129:89–105.
41. McCann ME, de Graaff JC, Dorris L, et al. GAS Consortium. Neurodevelopmental outcome at 5 years of age after general anaesthesia or awake-regional anaesthesia in infancy (GAS): an international, multicentre, randomised, controlled equivalence trial. *Lancet*. 2019;393(10172):664–677.

Stroke and translational research

9

Chandril Chugh
MAX Super Speciality Hospital, New Delhi, India

Stroke is one of the leading causes of morbidity and mortality worldwide. As per WHO, stroke is defined as a rapidly progressing neurological deficit that is vascular in origin.[1] Stroke is classified broadly into three categories: ischemic stroke, hemorrhagic stroke, and subarachnoid hemorrhage. Ischemic stroke occurs due to blockage of blood vessel which limits the blood supply to the brain whereas hemorrhagic stroke occurs due to rupture of blood vessel leading spillage of blood in the intracranial cavity.[2] Depending on the site of blood spillage the hemorrhagic stroke could be classified as intracerebral hemorrhage or subarachnoid hemorrhage. Approximately 60%–80% of all strokes are ischemic.

Over the last two decades, stroke mortality and morbidity have reduced considerably as intravenous thrombolysis and mechanical thrombectomy have been accepted as reliable modes of treatment. Although we still lack any consensus on treating chronic stroke patients as well as long-term recovery, fields of intracerebral hemorrhage and subarachnoid hemorrhage have seen some improvement due to improved understanding of principles of neurocritical care and use of modern endovascular techniques; however, the hope of permanent and long-term cure remains elusive.

Basic science and translational research have provided the clinical researchers with a variety of new stroke models, which have enhanced the understanding of stroke physiology and development of potential therapeutics that may help in recovery after stroke. Animal models have increased our understanding of time-based neuronal injury and potential for neurogenesis and neuroplasticity. Translational research can aid in direct and fast application of relevant findings into clinical research and further refine the treatment for stroke.

Stroke is a highly complex disease, which depends on a variety of both modifiable and nonmodifiable variables. Variables such as heterogeneity of vascular pathologies, patient demographics, comorbid conditions, and lack of validated biomarkers to stratify patient populations, limit the ability of stroke clinical trials to detect a treatment effect.[3] In the past, preclinical research has lacked the support and direction to provide clinicians with feasible and effective modes of treatment.

National Institute of Neurological Disorders and Stroke organized a workshop on Translational Stroke Research: Vision and Opportunities, in Bethesda, Maryland, in November 2016, to address the gaps in translational research. Development and manufacture of promising drugs, simplifying and justifying the use of animal models, designing studies with diverse demographics and patient

Perioperative Neuroscience. DOI: https://doi.org/10.1016/B978-0-323-91003-3.00012-X

factors and collaboration between scientists and clinicians were the core issues discussed. It was emphasized that for translation research to be successful study design should be standardized and the results be reproducible. Although a variety of animal models have been used in translational research, there needs to be clarity and reason behind the choice of the model as well as the endpoints. If there are multiple end points to be measured, the investigator should indicate how the study was powered and for what specific end point. With proper control and treatment groups, choice of statistical methods used in the analysis and interpretation of results becomes important.

Removal of study bias is another important aspect of translational research. Blinding and complete randomization of the study is recommended and concealment to be maintained till the results have been analyzed and the study completed. Both positive and negative results are significant and publishing them bears significance for future endeavors.[4] It was recommended and now preferred to reproduce the bench research findings in another laboratory and possibly in another species for validation. As this may be difficult and time consuming the idea of having multicenter translational research centers was proposed.[5]

Animal models for ischemic stroke

A variety of models have been used to study ischemic stroke in animals. Vascular access is usually done through craniectomy, and vessel occlusion can be achieved by either clipping, ligating, or using dyes such as Rose Bengal dye with light.[6–8] These models can be used in a variety of species and hence the results can be reproduced across species. Use of larger animals like dogs and pigs can be helpful due to the gyrencephalic cortices that have gray to white matter ratios that more closely mimic the human brain. It is easier to perform MRI scans and physiological monitoring, blood pressure and gaseous indices in these animals as well. In contrast, small animals offer the advantage of being far more cost effective to obtain and maintain and have the additional advantage of being genetically homogeneous and modifiable, specifically mice.[9]

Most common method of producing ischemia is to occlude the middle cerebral artery by using a nylon thread. The thread is introduced through the extracranial artery and pushed to the origin of the middle cerebral artery to achieve occlusion.[10] In certain cases, in larger animals balloons have also been used. Occlusion is confirmed through laser doppler flowmetry.[11] The thread model is useful as once the thread is removed reperfusion is achieved thus it enables the study of time-based injury, reperfusion injury and the mechanism of penumbra which closely replicates the physiology in the human brain. This model has helped tremendously in our understanding of excitotoxicity, glutaminergic toxicity, programmed cell death (apoptosis), oxidative stress, vascular damage and injury to the blood–brain barrier and inflammatory responses.[12] The drawback of this animal model is that doesn't mimic the thromboembolic stroke in humans. A large portion of ischemic stroke is thromboembolic in etiology. To understand this pathology embolic models have been developed which use catheter based injection of the clot in the middle cerebral artery. This can either achieved by using thrombin or by using arterial blood which clots by itself. As thrombin clots are very resistant to tissue plasminogen activator unmodified arterial blood is more widely used.[13,14] The major drawback to the embolic stroke models is the variability of the rate and timing of reperfusion and the lack of experimental control regarding the exact duration of ischemia. The use of autologous blood clots in these embolic models are currently the best model of cardioembolic strokes. However, this is not an optimal experimental model to assess complex temporal mechanisms. The intraluminal model gives good temporal control but lacks the specificity to human clinical condition. For any research to be successful, there has to be a hybrid of both the above approaches to reach a better understanding of ischemic stroke and time-dependent neuronal injury.[9]

Model for stroke risk factors

Stroke is a highly complex disease, and there are multiple risk factors ranging from age, sex, ethnicity to hypertension, obesity, diabetes and smoking. There are genetic and hereditary causes of stroke as well. These risk factors are grouped into either modifiable or nonmodifiable. The animal models designed to understand the risk for stroke need to concentrate on both these groups in order to reach any worthy conclusion.

There has been much debate over using ageing animals for study as stroke commonly affects the ageing brain that has often not been the case as that approach is expensive and time consuming. Some studies which have followed the protocol of ageing animals, have observed that ageing brains have larger stroke as compared to the younger ones in experimental stroke models. They have also observed that ageing brain responds differently to neuroprotectant drugs as compared to the younger counterparts.[15–17] Aged animals are more prone to oxidative stress, inflammation and vascular changes that more closely mimic the human population. Animal models that take into account the age and comorbidities albeit expensive are needed for any fruitful and meaningful research. Not only does the translational research based on such models should be encouraged unlike the past, rather they should be fast tracked. Thus, the progression in the translational research community must embrace the complexities of diverging signaling cascades and the various health/genetic factors, which may all contribute to more reliable models of the human condition.

Hyperglycemia and obesity

Diabetes Mellitus is one of the most important risk factors for stroke. Not only deranged blood sugar increases the risk of stroke it also impairs recovery after stroke and can cause other deadly complications such as myocardial infarction and kidney damage, both of which again increase the risk of stroke by themselves.[18] Obesity, similarly is a risk factor for stroke and increases the risk for diabetes and hypertension as well. Hyperglycemia has been shown to increase neuronal injury in ischemic stroke in many clinical studies. Not only does hyperglycemia cause neuronal death it also impairs neuroplasticity and repair. Impaired glucose tolerance has been associated with delayed functional recovery and poor outcomes. Numerous animal models are available for diabetes like toxin-induced Type 1 diabetes (alloxan and streptozotocin), as well as genetic animal models of Type 2 diabetes (db/db mouse, nonobese diabetic mouse and Zucker diabetic fatty rat).[19] Animal models of hyperglycemia and diabetes closely resemble the human response. In fact, both obesity and diabetes cause similar changes in terms of increased ischemic sensitivity. Hyperglycemia is associated with larger infarct core in animals (rats, dogs, rabbits) subjected to ischemia.[20,21] Mice with genetically induced diabetes also have larger ischemic cores. Multiple studies have shown that acidosis[22,23] and microvascular changes lead to increased damage in diabetic animals whereas in other models damage to the blood–brain barrier has been observed.[24,25]

There have been several studies, which have implicated other pathogenic mechanisms in stroke with diabetes such as increased inflammatory infiltration,[26,27] increased oxidative stress and MMP-9 activation,[24] impaired nitric oxide signaling[28] and vascular remodeling, which lead to reduced elasticity and altered resistance in diabetic animals.[29,30] Similar observations have also been made in humans.[31] Use of Insulin to reduce blood sugar has shown good results and reduction in ischemic injury. This positive effect has been attributed to normalization of endothelial nitric oxide signaling after insulin therapy.[32–34] Interestingly, similar effect of insulin was also observed in normoglycemic controls. Neuroprotective effects of insulin have been attributed to pathways downstream of brain insulin receptors which include PI3K/AKT interaction[35,36] which leads to enhanced synaptic plasticity, decreased excitability[37–39] and improved cerebral blood flow.[36]

Interaction of blood sugar with ischemic neurons is very complex. With ever increasing burden of diabetes in humans and the advent of newer oral hypoglycemic drugs our approach to basic clinical research should be more inclusive and aggressive. Future animal models will need to incorporate various drugs and their effect on animal models in order to increase our knowledge about this interplay of glucose and neurons, which to us, still lurks in the dark.

Hypertension

High blood pressure is the most common risk factor in stroke. Its prevalence among the stroke patients is anywhere from 35% to 50%. Aggressive control of blood pressure has been shown to reduce the risk of stroke. Animal models based around the study of hypertension have been helpful in our understanding of how high blood pressure affects the ischemic region, one such animal model is selectively bred spontaneously hypertensive rat (SHR). These rats are more prone to developing hypertension and consequently stroke.[39] Prolonged exposure to high blood pressure produces changes in vasculature, inflammatory reactions and chronic oxidative damage.[40–43] In rat model antihypertensive drugs seem to reduce ischemic injury when given prior to stroke as compared to post stroke.

Although there have been very few studies that investigate the long-term effect of antihypertensive drugs on cerebral ischemia, some other models have been used across specialities to study hypertension and its effects, such as the Dahl salt sensitive rat which develops hypertension on a high salt diet and the renal artery clipping model where hypertension is induced by clipping the renal artery.[44,45] In both these models the experimental stroke caused more injury with high blood pressure. In various studies on neuroprotection, it was observed that animals with higher blood pressure did worse than the controls with normal blood pressure. It is hypothesized that hypertension causes increased expression of the enzyme NADPH oxidase, which appears to be the primary source of reactive oxygen species in vasculature and may lead to vascular damage and nonreactivity.[46,47] Chronic hypertension may lead to oxidative stress which has been implicated in causing genomic changes that decrease the response to neuroprotectant drugs.[48–51] These genomic changes have been linked to altered metabolism, oxygen handling, excitability, blood–brain barrier function and inflammation. It has also been shown that SHR rats are more prone to ischemia and hypoxia induced apoptosis in invitro studies.

Animal studies based on the effects of hypertension are limited by unavailability of models which study chronic hypertension or controlled hypertension. To resemble the human population animal models which are based on controlled hypertension might seem more reasonable and more representative of the current scenario where most patients are on some kind of antihypertensive drug. Unfortunately, the interaction between chronic normalization of blood pressure in hypertensive animals and ischemic stroke outcome has not been reported. Ideally the animal studies would include ageing and hypertensive samples but, that can be time as well as resource consuming.

Gender

Stroke rates are higher in men irrespective of ethnicity and in both sexes stroke rate increases with age. In preclinical studies differences in sex have been studied with much interest. In one such study Hall et al. reported that female gebrils tend to have less neuronal damage as compared to male counterparts after stroke.[52] In rats too, females fared better and had less neuronal damage. Neuroprotection in females seems to diminish in studies where ovariectomized rats were used and inversely, neuroprotection was seen if the female rats were given estrogen before inducing stroke, suggesting the protective effect of female sex hormones.[53–55] However, the beneficial effects of estrogen in stroke seem to be time sensitive as its neuroprotective and antiinflammatory effects were lost with delayed estrogen replacement after ovariectomy or when given to aged animals.[56–58] There also seems to be a temporal association in acute stroke with early ($<$50 min) estrogen therapy proving to be more beneficial in limiting the infarct. It has been demonstrated that male animals benefited from acute postischemic estrogen treatment resulting in improved early reperfusion of CBF and attenuated brain damage.[59,60] Even in female SHR rats subjected to ischemic stroke during metestrus had a larger brain damage than rats that were exposed to ischemia and reperfusion during proestrus when estrogen levels are high.[61] Some of estrogen's neuroprotective effects have been linked to its antiinflammatory, antiapoptotic, and vasodilatory properties.[62]

Although protective effects are more pronounced in younger female animals similar effect has not been seen in ageing female animal population. Manwani et al observed that middle aged female mice had greater brain damage compared to middle aged males and young females.[63]

Neuronal protection

Acute neurological injury in stroke results due to massive calcium influx and glutamate excitotoxicity. In the past, many studies have been conducted to inhibit these pathways; however, a breakthrough is awaited. Multiple therapies targeting the glutamate release have been studied. Most promising were NMDA receptor antagonists MK-801, CGS-19755 and Gavestinel. All advanced to and failed phase III clinical trials due to poor side effect profiles, poor pharmacokinetics and limited brain penetration.[64] Another therapy designed to target the NRB2 receptor [NPS1506 and Traxoprodil (CP-101606)] was also unsuccessful. One of the reasons behind failure of these therapies has been hyperacute release of glutamate during ischemia which limits the clinical use of these drugs. AMPA receptor antagonists have also been studied however side effects like kidney toxicity, sedation, memory and visual impairment limited further development. Talampanel, a noncompetitive AMPA receptor antagonist, 2,3-benzodiazepine derivative has shown a good side effect profile unfortunately without much clinical benefit.[65]

Free radical scavengers

Another class of drugs is based on the principle of limiting free radical damage. Unfortunately, this approach too has been met with setbacks. A novel therapeutic which showed promise was nitrone disodium 2,4-disulphophenyl-N-tert-butylnitrone (NXY059). It was studied in multiple rodent and nonhuman primate models of stroke but was eventually found to be inadequate in modulating receptors of ischemic injury.[66–68] Edaravone, another free radical scavenger was studied in rat models with intracerebral hemorrhage and was found to be beneficial as it reduced the edema and neuronal injury. It acts against nitric oxide and reactive oxygen species release by activated microglia.[68,69] Edaravone was approved in Japan.

Phospholipase A2(PLA2) levels increase in acute stroke and lead to release of arachidonic acid, which releases free radicals through cyclooxygenase pathway (COX1/2).[70] Three kinds of PLA2 are known: Ca2þ-dependent secretory PLA2s (sPLA2s), Ca2þ-dependent cytosolic PLA2s (cPLA2s) and Ca2þ-independent cytosolic PLA2s (iPLA2s). 12-epi-scalaradial, a sPLA2 selective inhibitor failed to show any benefit in rat stroke model however less neuronal injury was observed in mice lacking cPLA2 expression.[71] Thus, it is possible that cPLA2 receptor inhibition may offer hope in further trials.[72] PLA2 is also involved in releasing lipid messengers that help in intercellular communication between microglia, neurons and endothelial cells. iPLA2 releases docosahexanoic acid in ischemic brain which is neuroprotective and antiinflammatory. Thus, selective therapies based on specific PLA2 expression may be helpful in neuroprotection in acute stroke.[73]

Citicholine, a phosphatidylcholine derivative, is known to decrease free radical formation by inhibiting free fatty acid metabolism and helps in recovery. It showed promising results in preclinical studies and was subjected to phase III trials based on that, which were also favorable. Citicholine was approved for post stroke recovery in parts of Europe based on these results.[74]

Another class of investigated drugs is acid sensing ion (ASIC) blockers. These drugs inhibit the lactic acid channels which are responsible massive calcium influx in ischemic brain. These have shown promising results in rat models.[75,76] It has been demonstrated that the pH of the ischemic brain allows full activation of ASIC1a, the dominant subtype in the central nervous system.[77] At this time, psalmotoxin venom is the only known subtype-selective blocker of ASIC1a.

Delayed neuronal injury

Delayed cell death in stroke occurs either due to Fas ligand based interaction or due to the activation of pro-apoptotic signaling protein Bax. Through both of these pathways caspase 3 is activated which leads to apoptosis or cell death.[78] Downregulation and inhibition of apoptotic genes is the focus of research to prevent neuronal loss. In some studies peptide based caspase receptors have shown promise in rodent models however, the results could not be replicated in clinical studies due to limited brain penetration.[78,79] Another focus has been PARP enzyme which leads to caspase independent cell death. PARP inhibitors have been shown to reduce infarct volume following occlusion of middle cerebral artery and improve neurological outcome in rats. PARP-1 inhibitor (ONO-2231) is currently being tested for acute ischemic stroke.[80,81] A different family of protein, cJun N-terminal kinase is responsible for cell death in ischemia, and its inhibition has shown good results in rodent models by promoting functional recovery.[82] Heat shock protein (hsp) have been studied due to their role in antiapoptosis mechanisms. Hsp70/90 increase the expression of antiapopotic gene bcl-2 which reduced inflammation and inhibit the expression of matrix metalloproteinases. Arimoclomol is one such drugs which aids in the expression of hsp70 and hsp 90, and has been shown to improve sensorimotor recovery after ischemia.[83]

Inflammation

Ischemia leads to infiltration of the tissue with neutrophils and lymphocytes, which secrete and upregulate cytokines and inflammatory markers. These proinflammatory compounds lead to edema and cell death. Src family nonreceptor protein tyrosine kinases have been shown to cause endothelial damage by phosphorylation of cell-cell adhesion molecules which leads to increased permeability and edema.[84] Studies focussing on Src inhibitors have shown encouraging results in rodent models of neuroprotection.[85] Interleukin-1 and tumor necrosis factor (TNF-alpha) are proinflammatory markers which cause neuronal damage after stroke. IL-1 is secreted by microglia, astrocytes and neurons and leads to inflammation through arachidonic acid pathway and NMDA receptor mediated glutamate excitotoxicity.[86] TNF levels are also increased in the cerebrospinal fluid in rodent stroke models. Inhibition of these mediators is helpful in reducing neuronal death. In a study, administering the IL-1 antagonist IL-1RA to human patients with stroke reduced the level of circulating inflammatory cells without any side effects.[87,88] Other molecules such as selectins, integrins and adhesion molecules also play an important role in inflammation by increasing the infiltration of neutrophils and polymorphonuclear lymphocytes at the site of ischemia. Inhibition of these molecules has been shown to reduce infarct size in rat models.[89] Intercellular adhesion molecule-1 or vascular cell adhesion molecule-1 levels in the CSF corelates with stroke severity and therapies directed towards them although theoretically promising have failed to show any clinical benefit.[90,91]

Regeneration in stroke and neuronal repair

Post ischemia, mainly two modes of repair have been observed, cell replacement and plasticity. After ischemic injury there is upregulation of neurogenesis and stem cell migration to the ischemic site. Animal models have shown cell migration to the ischemic zone and neurogenesis by precursor cells from subventricular zone in cortex and striatum.[92,93] Although most of the neuroblasts are lost over time suggesting that the brain environment may not be optimal for their survival, some of them form synapses with the preexisting neurons in the ischemic zone. Whether this leads to recovery and improvement is difficult to prove.[94,95] Migration of neuroblasts forms the basis of stem cell research in stroke however, a number of challenges have been detrimental in wider acceptance of this approach.

There is growing evidence based on imaging techniques that areas close to ischemic tissue form synapses and try to compensate for the neuronal loss, this is called plasticity. Cortical sensorimotor and premotor areas exhibit great deal of plasticity and appear to compensate for lost synapses by forming new ones and help in functional recovery.[96–98] It has been observed through electron microscopy that early neuronal loss is followed by regeneration of axons, dendrites and synapses in penumbra as well as the core region of ischemia. Extensive remodeling in peri-infarct region has been shown to lead to at least partial functional recovery.[99,100]

Endogenous repair mechanisms lead to partial recovery at best. There is growing evidence that physiotherapy and cognitive therapy may help and promote these processes leading to further recovery. The holy grail of neuronal regeneration and significant functional recovery may very well lie in the pharmacological advancement and development of therapeutics that will support the endogenous repair modalities.

The idea of promoting endogenous processes of repair led to the development of drugs that promote plasticity by downregulating inhibitors. In normal conditions dendritic and axonal proteins limit plasticity. Neutralization of myelin receptor Nogo receptor 1 (NgR1) using NgR1-Fc fusion protein targets multiple myelin-based proteins involved in plasticity. It has been shown that NgR1 based therapy administered within 7 days post ischemia led to significant motor improvement and axonal plasticity at 70 days without decreasing lesion volume. This has been replicated in NgR-1 negative mice where significant axonal growth and synapse formation was observed one month after the injury.[101] Specific antibodies against myelin associated glycoprotein led to partial recovery in animal models.[102] Inhibition of Nogo-A, which is a NgR-1 independent pathway by specific antibodies also led to improvement in motor function when treatment was given within seven days of insult. This result however, was more pronounced in younger rats, the older rats took longer to show motor improvement.[103,104] Rodent models of Nogo A neutralization or NogoA/B ablation observed axonal sprouting to deafferented targets, increased dendritic arborization in contralateral cortex suggesting the role of contralateral non injured cortex in aiding functional recovery.[105,106] In addition to NogoA and NgR-1, other factors like Rho-kinase and cAMP also appear to be regenerative in longer therapeutic windows.[107,108] Future development of therapeutics will most likely target these factors to help in neuronal plasticity.

Growth promoters

Brain-derived neurotrophic factor (BDNF) and basic fibroblast growth factor (bFGF) are two known neuronal growth promoting factors found in the brain, which lead to improved plasticity and axonal growth. During ischemia, the levels of these factors increase however, exogenous supplementation has been found to be beneficial nevertheless.[109] Supplementation with bFGF within a day of stroke has been shown to improve sensorimotor recovery without affecting the infarct volume. bFGF leads to axonal sprouting, is antiapoptotic and causes neuronal stem cell proliferation and differentiation. Although the animal trials were promising the human trials could not replicate the results, still some benefit was observed in patients who received infusion.[110,111] Prolonged infusion of BDNF (5 days) in rodent stroke models led to significant sensorimotor improvement. BDNF works by stimulation of neuronal precursor cells in striatum and has also been shown to improve neuroangiogenesis in the infarcted tissue.[112,113] BDNF binds to both TrkB receptor and to the p75 neurotrophin receptor. Agonistic molecules directed to these receptors have been useful in potentiating neuropoietic effects of BDNF while inhibiting apoptosis mediated by p 75 receptor.[114]

Vascular endothelial growth factor (VEGF) is rapidly upregulated in ischemia and promotes vascular leakage in the early phase followed by neuroangiogenesis and recovery in the late phase. VEGF is associated with striatal neurogenesis when administered in the ventricles or when it is upregulated at the time of ischemia.[89,115] Recombinant human erythropoietin (rEPO) exerts its neuroprotective effects by neurogenesis and proliferation of endothelial cells.[116] Both VEGF and rEPO lead to neurovascular repair by endothelial cell proliferation. Increased endothelial cell proliferation has been associated with improved outcome in acute ischemic stroke.[117,118] Finally, hematopoetic growth factors, including granulocyte-colony stimulating factor aid in neurogenesis which has resulted in functional improvement in rat model.[119]

Other drugs

Drugs like minocycline have also been studied in acute stroke recovery. It has been shown to promote neurogenesis and survival of hippocampal neurons following ischemia. Minocycline has a neuroprotective as well as an antiinflammatory activity which has been shown to improve outcome in a long therapeutic window human stroke trial.[120]

Immunophilin ligands are another class of studied therapeutics. Immunosuppressive drugs like FK506 and rapamycin showed limited efficacy in ischemia models.[121,122] Non immunosuppressive ligands like GPI-1046 and V10367 also showed some neuroprotective effects in animal models.[123] Another compound ILS-920 has shown promising neuroprotective effects by decreasing infarct volume and reducing tissue inflammation.[124]

A number of other compounds ranging from hormones like estrogen and drugs like statins have been studied with variable results.

Conclusion

Basic research is the bed rock of meaningful clinical research. Encouraging results in the labs can be translated into useful clinical findings, which can be used to help patients with ischemic stroke. Translational research focus should be on high yield and most promising areas of basic science research to help fast track their application in the clinical scenario for maximum benefit. Allocation of resources should take into account the clinical utility, economic feasibility, and commercial viability of the study drug or therapeutic. As our knowledge about the pathophysiology of stroke increases, more and more avenues of research will open up. It will be then up to the scientists and the policy makers to choose the area of focus not only based on urgency but also on the likelihood of clinical success and economic costs.

References

1. Hatano S. Experience from a multicentre stroke register: a preliminary report. *Bull World Health Organ.* 1976;54(5):541–553.
2. Caplan LR. *Basic pathology, anatomy, and pathophysiology of stroke. Caplan's Stroke: A Clinical Approach.* 4th ed Philadelphia: Saunders Elsevier; 2009:22.

3. Bosetti F, Koenig JI, Ayata C, et al. Translational stroke research: vision and opportunities. *Stroke*. 2017;48(9):2632–2637.
4. Lapchak PA. Recommendations and practices to optimize stroke therapy: developing effective translational research programs. *Stroke*. 2013;44(3):841–843.
5. Dirnagl U, Fisher M. International, multicenter randomized preclinical trials in translational stroke research: it's time to act.
6. Tamura A, Graham DI, Mcculloch J, Teasdale GM. Focal cerebral ischaemia in the rat: 1. Description of technique and early neuropathological consequences following middle cerebral artery occlusion. *J Cereb Blood Flow Metab*. 1981;1(1):53–60.
7. Tamura A, Asano T, Sano K, Tsumagari T, Nakajima A. Protection from cerebral ischemia by a new imidazole derivative (Y-9179) and pentobarbital. A comparative study in chronic middle cerebral artery occlusion in cats. *Stroke*. 1979;10(2):126–134.
8. Markgraf CG, Kraydieh S, Prado R, Watson BD, Dietrich WD, Ginsberg MD. Comparative histopathologic consequences of photothrombotic occlusion of the distal middle cerebral artery in Sprague-Dawley and Wistar rats. *Stroke*. 1993;24(2):286–292.
9. Herson PS, Traystman RJ. Animal models of stroke: translational potential at present and in 2050. *Future Neurol*. 2014;9(5):541–551.
10. Longa EZ, Weinstein PR, Carlson S, Cummins R. Reversible middle cerebral artery occlusion without craniectomy in rats. *Stroke*. 1989;20(1):84–91.
11. Gao H, Liu Y, Lu S, Xiang B, Wang C. A reversible middle cerebral artery occlusion model using intraluminal balloon technique in monkeys. *J Stroke Cereb Dis*. 2006;15(5):202–208.
12. Moskowitz MA, Lo EH, Iadecola C. The science of stroke: mechanisms in search of treatments. *Neuron*. 2010;67(2):181–198.
13. Kudo M, Aoyama A, Ichimori S, Fukunaga N. An animal model of cerebral infarction. Homologous blood clot emboli in rats. *Stroke*. 1982;13(4):505–508.
14. Niessen F, Hilger T, Hoehn M, Hossmann KA. Differences in clot preparation determine outcome of recombinant tissue plasminogen activator treatment in experimental thromboembolic stroke. *Stroke*. 2003;34(8):2019–2024.
15. Badan I, Buchhold B, Hamm A, et al. Accelerated glial reactivity to stroke in aged rats correlates with reduced functional recovery. *J Cereb Blood Flow Metab*. 2003;23(7):845–854.
16. Ay H, Koroshetz WJ, Vangel M, et al. Conversion of ischemic brain tissue into infarction increases with age. *Stroke*. 2005;36(12):2632–2636.
17. Popa-Wagner A, Buga AM, Kokaia Z. Perturbed cellular response to brain injury during aging. *Ageing Res Rev*. 2011;10(1):71–79.
18. Kaarisalo MM, Raiha I, Sivenius J, et al. Diabetes worsens the outcome of acute ischemic stroke. *Diabetes Res Clin Pract*. 2005;69(3):293–298.
19. King AJ. The use of animal models in diabetes research. *Br J Pharmacol*. 2012;166(3):877–894.
20. Liu L, Wang Z, Wang X, et al. Comparison of two rat models of cerebral ischemia under hyperglycemic conditions. *Microsurgery*. 2007;27(4):258–262.
21. Huang NC, Wei J, Quast MJ. A comparison of the early development of ischemic brain damage in normoglycemic and hyperglycemic rats using magnetic resonance imaging. *Exp Brain Res*. 1996;109(1):33–42.
22. Siesjo BK, Katsura KI, Kristian T, Li PA, Siesjo P. Molecular mechanisms of acidosis-mediated damage. *Acta Neurochir Suppl*. 1996;66:8–14.
23. Kawai N, Keep RF, Betz AL, Nagao S. Hyperglycemia induces progressive changes in the cerebral microvasculature and blood-brain barrier transport during focal cerebral ischemia. *Acta Neurochir Suppl*. 1998;71:219–221.

24. Kamada H, Yu F, Nito C, Chan PH. Influence of hyperglycemia on oxidative stress and matrix metalloproteinase-9 activation after focal cerebral ischemia/reperfusion in rats: relation to blood–brain barrier dysfunction. *Stroke*. 2007;38(3):1044–1049.
25. Kittaka M, Wang L, Sun N, et al. Brain capillary tissue plasminogen activator in a diabetes stroke model. *Stroke*. 1996;27(4):712–719.
26. Panes J, Kurose I, Rodriguez-Vaca D, et al. Diabetes exacerbates inflammatory responses to ischemia–reperfusion. *Circulation*. 1996;93(1):161–167.
27. Ding C, He Q, Li PA. Diabetes increases expression of ICAM after a brief period of cerebral ischemia. *J Neuroimmunol*. 2005;161(1–2):61–67.
28. Li Q, Atochin D, Kashiwagi S, et al. Deficient eNOS phosphorylation is a mechanism for diabetic vascular dysfunction contributing to increased stroke size. *Stroke*. 2013;44(11):3183–3188.
29. Schofield I, Malik R, Izzard A, Austin C, Heagerty A. Vascular structural and functional changes in Type 2 diabetes mellitus: evidence for the roles of abnormal myogenic responsiveness and dyslipidemia. *Circulation*. 2002;106(24):3037–3043.
30. Onodera H, Oshio K, Uchida M, Tanaka Y, Hashimoto T. Analysis of intracranial pressure pulse waveform and brain capillary morphology in Type 2 diabetes mellitus rats. *Brain Res*. 2012;1460:73–77.
31. Hou Q, Zuo Z, Michel P, et al. Influence of chronic hyperglycemia on cerebral microvascular remodeling: an in vivo study using perfusion computed tomography in acute ischemic stroke patients. *Stroke*. 2013;44(12):3557–3560.
32. Warner DS, Gionet TX, Todd MM, Mcallister AM. Insulin-induced normoglycemia improves ischemic outcome in hyperglycemic rats. *Stroke*. 1992;23(12):1775–1780.
33. Hamilton MG, Tranmer BI, Auer RN. Insulin reduction of cerebral infarction due to transient focal ischemia. *J Neurosurg*. 1995;82(2):262–268.
34. Huang SS, Lu YJ, Huang JP, Wu YT, Day YJ, Hung LM. The essential role of endothelial nitric oxide synthase activation in insulinmediated neuroprotection against ischemic stroke in diabetes. *J Vasc Surg*. 2013;59(2):483–491.
35. Philpott KL, Mccarthy MJ, Klippel A, Rubin LL. Activated phosphatidylinositol 3-kinase and Akt kinase promote survival of superior cervical neurons. *J Cell Biol*. 1997;139(3):809–815.
36. Sanderson TH, Kumar R, Murariu-Dobrin AC, Page AB, Krause GS, Sullivan JM. Insulin activates the PI3K–Akt survival pathway in vulnerable neurons following global brain ischemia. *Neurol Res*. 2009;31(9):947–958.
37. Mielke JG, Wang YT. Insulin exerts neuroprotection by counteracting the decrease in cell-surface GABA receptors following oxygen-glucose deprivation in cultured cortical neurons. *J Neurochem*. 2005;92(1):103–113.
38. Zhao WQ, Alkon DL. Role of insulin and insulin receptor in learning and memory. *Mol Cell Endocrinol*. 2001;177(1–2):125–134.
39. Mckay MK, Hester RL. Role of nitric oxide, adenosine, and ATP-sensitive potassium channels in insulin-induced vasodilation. *Hypertension*. 1996;28(2):202–208.
40. Jimenez-Altayo F, Martin A, Rojas S, et al. Transient middle cerebral artery occlusion causes different structural, mechanical, and myogenic alterations in normotensive and hypertensive rats. *Am J Physiol Heart Circ Physiol*. 2007;293(1):H628–H635.
41. Pires PW, Dams Ramos CM, Matin N, Dorrance AM. The effects of hypertension on the cerebral circulation. *Am J Physiol Heart Circ Physiol*. 2013;304(12):H1598–H1614.
42. Harrison DG, Marvar PJ, Titze JM. Angiotensin II AT1 receptor blockade abolishes brain microvascular inflammation and heat shock protein responses in hypertensive ratsVascular inflammatory cells in hypertension. *Front Physiol*. 2012;3:128. 42 Zhou J, Ando H, Macova M, Dou J, Saavedra JM *J Cereb Blood Flow Metab*. 2005;25(7):878–886.

43. Touyz RM, Briones AM. Reactive oxygen species and vascular biology: implications in human hypertension. *Hypertens Res*. 2011;34(1):5–14.
44. Bright R, Steinberg GK, Mochly-Rosen D. DeltaPKC mediates microcerebrovascular dysfunction in acute ischemia and in chronic hypertensive stress in vivo. *Brain Res*. 2007;1144:146–155.
45. Fujishima M, Onoyama K, Oniki H, Ogata J, Omae T. Effects of acute hypertension on brain metabolism in normotensive, renovascular hypertensive and spontaneously hypertensive rats. *Stroke*. 1978;9(4):349–353.
46. Macleod MR, Van Der Worp HB, Sena ES, Howells DW, Dirnagl U, Donnan GA. Evidence for the efficacy of NXY-059 in experimental focal cerebral ischaemia is confounded by study quality. *Stroke*. 2008;39(10):2824–2829.
47. Briones AM, Touyz RM. Oxidative stress and hypertension: current concepts. *Curr Hypertens Rep*. 2010;12(2):135–142.
48. Ritz MF, Grond-Ginsbach C, Engelter S, Lyrer P. Gene expression suggests spontaneously hypertensive rats may have altered metabolism and reduced hypoxic tolerance. *Curr Neurovasc Res*. 2012;9(1):10–19.
49. Qiu J, Hong Q, Chen RH, et al. Gene expression profiles in the prefrontal cortex of SHR rats by cDNA microarrays. *Mol Biol Rep*. 2010;37(4):1733–1740.
50. Ferrari MF, Reis EM, Matsumoto JP, Fior-Chadi DR. Gene expression profiling of cultured cells from brainstem of newborn spontaneously hypertensive and Wistar Kyoto rats. *Cell Mol Neurobiol*. 2009;29(3):287–308.
51. Ishida H, Takemori K, Dote K, Ito H. Expression of glucose transporter-1 and aquaporin-4 in the cerebral cortex of stroke-prone spontaneously hypertensive rats in relation to the blood–brain barrier function. *Am J Hypertens*. 2006;19(1):33–39.
52. Hall ED, Pazara KE, Linseman KL. Sex differences in postischemic neuronal necrosis in gerbils. *J Cereb Blood Flow Metab*. 1991;11:292–298.
53. Alkayed NJ, Harukuni I, Kimes AS, London ED, Traystman RJ, Hurn PD. Gender-linked brain injury in experimental stroke. *Stroke*. 1998;29:159–165. discussion 66.
54. Dubal DB, Kashon ML, Pettigrew LC, et al. Estradiol protects against ischemic injury. *J Cereb Blood Flow Metab*. 1998;18:1253–1258.
55. Simpkins JW, Rajakumar G, Zhang YQ, et al. Estrogens may reduce mortality and ischemic damage caused by middle cerebral artery occlusion in the female rat. *J Neurosurg*. 1997;87:724–730.
56. Suzuki S, Brown CM, Dela Cruz CD, Yang E, Bridwell DA, Wise PM. Timing of estrogen therapy after ovariectomy dictates the efficacy of its neuroprotective and antiinflammatory actions. *Proc Natl Acad Sci U S A*. 2007;104:6013–6018.
57. Cai M, Ma YL, Qin P, et al. The loss of estrogen efficacy against cerebral ischemia in aged postmenopausal female mice. *Neurosci Lett*. 2014;558:115–119.
58. Liu F, Benashski SE, Xu Y, Siegel M, McCullough LD. Effects of chronic and acute oestrogen replacement therapy in aged animals after experimental stroke. *J Neuroendocrinol*. 2012;24:319–330.
59. Yang SH, Shi J, Day AL, Simpkins JW. Estradiol exerts neuroprotective effects when administered after ischemic insult. *Stroke*. 2000;31:745–749. discussion 9–50.
60. McCullough LD, Alkayed NJ, Traystman RJ, Williams MJ, Hurn PD. Postischemic estrogen reduces hypoperfusion and secondary.
61. Carswell HV, Dominiczak AF, Macrae IM. Estrogen status affects sensitivity to focal cerebral ischemia in stroke-prone spontaneously hypertensive rats. *Am J Physiol Heart Circ Physiol*. 2000;278:H290–H294.

62. Ahnstedt H, McCullough LD, Cipolla MJ. The importance of considering sex differences in translational stroke research. *Transl Stroke Res*. 2016;7(4):261–273.
63. Manwani B, Liu F, Scranton V, Hammond MD, Sansing LH, McCullough LD. Differential effects of aging and sex on stroke induced inflammation across the lifespan. *Exp Neurol*. 2013;249:120–131.
64. Hoyte L, Barber PA, Buchan AM, Hill MD. The rise and fall of NMDA antagonists for ischemic stroke. *Curr Mol Med*. 2004;4:131–136.
65. Howes JF, Bell C. Talampanel. *Neurotherapeutics*. 2007;4:126–129.
66. Kuroda S, Tsuchidate R, Smith ML, Maples KR, Siesjo BK. Neuroprotective effects of a novel nitrone, NXY-059, after transient focal cerebral ischemia in the rat. *J Cereb Blood Flow Metab*. 1999;19:778–787.
67. Marshall JW, Duffin KJ, Green AR, Ridley RM. NXY-059, a free radical– trapping agent, substantially lessens the functional disability resulting from cerebral ischemia in a primate species. *Stroke*. 2001;32:190–198.
68. Feuerstein GZ, Zaleska MM, Krams M, et al. Missing steps in the STAIR case: a translational medicine perspective on the development of NXY-059 for treatment of acute ischemic stroke. *J Cereb Blood Flow Metab*. 2008;28:217–219.
69. Banno M, Mizuno T, Kato H, et al. The radical scavenger edaravone prevents oxidative neurotoxicity induced by peroxynitrite and activated microglia. *Neuropharmacology*. 2005;48:283–290.
70. Nakamura T, Kuroda Y, Yamashita S, et al. Edaravone attenuates brain edema and neurologic deficits in a rat model of acute intracerebral hemorrhage. *Stroke*. 2008;39:463–469.
71. Torregrosa G, Perez-Asensio FJ, Burguete MC, Castello-Ruiz M, Salom JB, Alborch E. Chronic intracerebroventricular delivery of the secretory phospholipase A2 inhibitor, 12-epi-scalaradial, does not improve outcome after focal cerebral ischemia-reperfusion in rats. *Exp Brain Res*. 2007;176:248–259.
72. Bonventre JV. Roles of phospholipases A2 in brain cell and tissue injury associated with ischemia and excitotoxicity. *J Lipid Mediat Cell Signal*. 1997;17:71–79 (corrected and republished from Bonventre JV. *J Lipid Mediat Cell Signal*. 1996;14: 15–23.
73. Green AR. Pharmacological approaches to acute ischaemic stroke: reperfusion certainly, neuroprotection possibly. *Br J Pharmacol*. 2008;153(suppl 1):S325–S338.
74. Adibhatla RM, Hatcher JF. Citicoline mechanisms and clinical efficacy in cerebral ischemia. *J Neurosci Res*. 2002;70:133–139.
75. Pignataro G, Simon RP, Xiong ZG. Prolonged activation of ASIC1a and the time window for neuroprotection in cerebral ischaemia. *Brain*. 2007;130:151–158.
76. Xiong ZG, Pignataro G, Li M, Chang SY, Simon RP. Acid-sensing ion channels (ASICs) as pharmacological targets for neurodegenerative diseases. *Curr Opin Pharmacol*. 2008;8:25–32.
77. Wemmie JA, Price MP, Welsh MJ. Acid-sensing ion channels: advances, questions and therapeutic opportunities. *Trends Neurosci*. 2006;29:578–586.
78. Green DR, Reed JC. Mitochondria and apoptosis. *Science*. 1998;281:1309–1312.
79. Li H, Colbourne F, Sun P, Zhao Z, Buchan AM, Iadecola C. Caspase inhibitors reduce neuronal injury after focal but not global cerebral ischemia in rats. *Stroke*. 2000;31:176–182.
80. Schulz JB, Weller M, Moskowitz MA. Caspases as treatment targets in stroke and neurodegenerative diseases. *Ann Neurol*. 1999;45:421–429.
81. Moroni F. Poly(ADP-ribose)polymerase 1 (PARP-1) and postischemic brain damage. *Curr Opin Pharmacol*. 2008;8:96–103.

82. Esneault E, Castagne V, Moser P, Bonny C, Bernaudin M. D-JNKi, a peptide inhibitor of c-Jun N-terminal kinase, promotes functional recovery after transient focal cerebral ischemia in rats. *Neuroscience*. 2008;152:308–320.
83. Brown IR. Heat shock proteins and protection of the nervous system. *Ann NY Acad Sci*. 2007;1113:147–158.
84. Eliceiri BP, Paul R, Schwartzberg PL, Hood JD, Leng J, Cheresh DA. Selective requirement for Src kinases during VEGF-induced angiogenesis and vascular permeability. *Mol Cell*. 1999;4:915–924.
85. Kago T, Takagi N, Date I, Takenaga Y, Takagi K, Takeo S. Cerebral ischemia enhances tyrosine phosphorylation of occludin in brain capillaries. *Biochem Biophys Res Commun*. 2006;339:1197–1203.
86. Simi A, Tsakiri N, Wang P, Rothwell NJ. Interleukin-1 and inflammatory neurodegeneration. *Biochem Soc Trans*. 2007;35:1122–1126.
87. Vila N, Castillo J, Davalos A, Chamorro A. Proinflammatory cytokines and early neurological worsening in ischemic stroke. *Stroke*. 2000;31:2325–2329.
88. Clark SR, McMahon CJ, Gueorguieva I, et al. Interleukin-1 receptor antagonist penetrates human brain at experimentally therapeutic concentrations. *J Cereb Blood Flow Metab*. 2008;28:387–394.
89. Wang Q, Tang XN, Yenari MA. The inflammatory response in stroke. *J Neuroimmunol*. 2007;184:53–68.
90. Ehrensperger E, Minuk J, Durcan L, et al. Predictive value of soluble intercellular adhesion molecule-1 for risk of ischemic events in individuals with cerebrovascular disease. *Cerebrovasc Dis*. 2005;20:456–462.
91. Simundic AM, Basic V, Topic E, et al. Soluble adhesion molecules in acute ischemic stroke. *Clin Invest Med Med Clin Exp*. 2004;27:86–92.
92. Arvidsson A, Collin T, Kirik D, Kokaia Z, Lindvall O. Neuronal replacement from endogenous precursors in the adult brain after stroke. *Nat Med*. 2002;8:963–970.
93. Magavi SS, Leavitt BR, Macklis JD. Induction of neurogenesis in the neocortex of adult mice. *Nature*. 2000;405:951–955.
94. Leker RR, Soldner F, Velasco I, Gavin DK, Androutsellis-Theotokis A, McKay RD. Long-lasting regeneration after ischemia in the cerebral cortex. *Stroke*. 2007;38:153–161.
95. Yamashita T, Ninomiya M, Hernandez Acosta P, et al. Subventricular zone-derived neuroblasts migrate and differentiate into mature neurons in the post-stroke adult striatum. *J Neurosci*. 2006;26:6627–6636.
96. Nudo RJ. Mechanisms for recovery of motor function following cortical damage. *Curr Opin Neurobiol*. 2006;16:638–644.
97. O'Shea J, Johansen-Berg H, Trief D, Gobel S, Rushworth MF. Functionally specific reorganization in human premotor cortex. *Neuron*. 2007;54:479–490.
98. Rossini PM, Altamura C, Ferreri F, et al. Neuroimaging experimental studies on brain plasticity in recovery from stroke. *Eura Medicophys*. 2007;43:241–254.
99. Brown CE, Li P, Boyd JD, Delaney KR, Murphy TH. Extensive turnover of dendritic spines and vascular remodeling in cortical tissues recovering from stroke. *J Neurosci*. 2007;27:4101–4109.
100. Brown CE, Murphy TH. Livin' on the edge: imaging dendritic spine turnover in the peri-infarct zone during ischemic stroke and recovery. *Neuroscientist*. 2008;14:139–146.
101. Lee JK, Kim JE, Sivula M, Strittmatter SM. Nogo receptor antagonism promotes stroke recovery by enhancing axonal plasticity. *J Neurosci*. 2004;24:6209–6217.

102. Irving EA, Vinson M, Rosin C, et al. Identification of neuroprotective properties of anti-MAG antibody: a novel approach for the treatment of stroke? *J Cereb Blood Flow Metab.* 2005;25:98–107.
103. Freund P, Schmidlin E, Wannier T, et al. Nogo-A-specific antibody treatment enhances sprouting and functional recovery after cervical lesion in adult primates. *Nat Med.* 2006;12:790–792.
104. Papadopoulos CM, Tsai SY, Alsbiei T, O'Brien TE, Schwab ME, Kartje GL. Functional recovery and neuroanatomical plasticity following middle cerebral artery occlusion and IN-1 antibody treatment in the adult rat. *Ann Neurol.* 2002;51:433–441.
105. Papadopoulos CM, Tsai SY, Cheatwood JL, et al. Dendritic plasticity in the adult rat following middle cerebral artery occlusion and Nogo-a neutralization. *Cereb Cortex.* 2006;16:529–536.
106. Eslamboli A, Grundy RI, Irving EA. Time-dependent increase in Nogo-A expression after focal cerebral ischemia in marmoset monkeys. *Neurosci Lett.* 2006;408:89–93.
107. Sasaki T, Kitagawa K, Omura-Matsuoka E, et al. The phosphodiesterase inhibitor rolipram promotes survival of newborn hippocampal neurons after ischemia. *Stroke.* 2007;38:1597–1605.
108. Satoh S, Toshima Y, Hitomi A, Ikegaki I, Seto M, Asano T. Wide therapeutic time window for Rho-kinase inhibition therapy in ischemic brain damage in a rat cerebral thrombosis model. *Brain Res.* 2008;1193:102–108.
109. Sugimori H, Speller H, Finklestein SP. Intravenous basic fibroblast growth factor produces a persistent reduction in infarct volume following permanent focal ischemia in rats. *Neurosci Lett.* 2001;300:13–16.
110. Ren JM, Finklestein SP. Growth factor treatment of stroke. *Curr Drug Targets CNS Neurol Disord.* 2005;4:121–125.
111. Berry D, Ren J, Kwan CP, Sietsma DK, Sasisekharan R, Finklestein SP. Dimeric fibroblast growth factor-2 enhances functional recovery after focal cerebral ischemia. *Restor Neurol Neurosci.* 2005;23:251–256.
112. Schabitz WR, Steigleder T, Cooper-Kuhn CM, et al. Intravenous brain-derived neurotrophic factor enhances poststroke sensorimotor recovery and stimulates neurogenesis. *Stroke.* 2007;38:2165–2172.
113. Kermani P, Hempstead B. Brain-derived neurotrophic factor: a newly described mediator of angiogenesis. *Trends Cardiovasc Med.* 2007;17:140–143.
114. Qian MD, Zhang J, Tan XY, Wood A, Gill D, Cho S. Novel agonist monoclonal antibodies activate TrkB receptors and demonstrate potent neurotrophic activities. *J Neurosci.* 2006;26:9394–9403.
115. Wang YQ, Guo X, Qiu MH, Feng XY, Sun FY. VEGF overexpression enhances striatal neurogenesis in brain of adult rat after a transient middle cerebral artery occlusion. *J Neurosci Res.* 2007;85:73–82.
116. Gonzalez FF, McQuillen P, Mu D, et al. Erythropoietin enhances long-term neuroprotection and neurogenesis in neonatal stroke. *Dev Neurosci.* 2007;29:321–330.
117. Sobrino T, Hurtado O, Moro MA, et al. The increase of circulating endothelial progenitor cells after acute ischemic stroke is associated with good outcome. *Stroke.* 2007;38:2759–2764.
118. Carmichael ST. Themes and strategies for studying the biology of stroke recovery in the poststroke epoch. *Stroke.* 2008;39:1380–1388.
119. Zhao LR, Singhal S, Duan WM, Mehta J, Kessler JA. Brain repair by hematopoietic growth factors in a rat model of stroke. *Stroke.* 2007;38:2584–2591.

120. Lampl Y, Boaz M, Gilad R, et al. Minocycline treatment in acute stroke: an open-label, evaluator-blinded study. *Neurology*. 2007;69:1404–1410.
121. Butcher SP, Henshall DC, Teramura Y, Iwasaki K, Sharkey J. Neuroprotective actions of FK506 in experimental stroke: in vivo evidence against an antiexcitotoxic mechanism. *J Neurosci*. 1997;17:6939–6946.
122. Sharkey J, Butcher SP. Immunophilins mediate the neuroprotective effects of FK506 in focal cerebral ischaemia. *Nature*. 1994;371:336–339.
123. Li F, Omori N, Hayashi T, et al. Protection against ischemic brain damage in rats by immunophilin ligand GPI1046. *J Neurosci Res*. 2004;76:383–389.
124. Ruan B, Pong K, Jow F, et al. Binding of rapamycin analogs to calcium channels and FKBP52 contributes to their neuroprotective activities. *Proc Natl Acad Sci USA*. 2008;105:33–38.

Traumatic brain injury and translational research: pharmacological and nonpharmacological perspectives

10

Denise Battaglini[1,2], *Chiara Robba*[1,3] *and Paolo Pelosi*[1,3]
[1]Anesthesia and Intensive Care, San Martino Policlinico Hospital, IRCCS for Oncology and Neuroscience, Genoa, Italy, [2]Department of Medicine, University of Barcelona (UB), Barcelona, Spain, [3]Department of Surgical Sciences and Integrated Diagnostics, University of Genoa, Genoa, Italy

Introduction

Traumatic brain injury (TBI) is an important cause of disability and death worldwide.[1] The primary brain damage is followed by a detrimental secondary brain injury if not appropriately managed.[2] Management of TBI has been changed over the years and novel therapeutic pharmacological and nonpharmacological strategies have been implemented.[2,3] During the management of TBI, a multimodal neuromonitoring strategy is always suggested when deemed available (Fig. 10.1).

Pathophysiology of traumatic brain injury

Patients with TBI are at higher risk of secondary brain injury, poor outcome, and long-term disability.[4] Pathophysiology of primary brain injury includes focal or diffuse damage, with possible brain hypoxic injury. Subsequently, glial and neuronal death can occur, followed by altered blood supply and damage of the blood–brain barrier (BBB) integrity.[5] Postinjury inflammation with the release of proinflammatory cytokines may contribute to vasogenic edema and damage of blood vessels.[6] The injured brain tissue is associated with inflammation, astrogliosis, proliferation, and hypertrophy of astrocytes acting to sequester the necrotic and inflammatory area from the rest of the brain tissue.[6,7] Other factors are involved, including excitotoxicity, mitochondrial dysfunction, oxidative stress, lipid peroxidation, axon degeneration, and apoptotic cell death.[8] Activated pathways and released enzymes after cytotoxicity and redox processes worsen the brain damage and all these mechanisms are mainly processed by mitochondria.[9] Around the brain lesion, Wallerian axonal degeneration with infiltration by macrophages, which activate proteolysis, excitotoxicity, and cause mitochondrial dysfunction can be observed.[10] The persistent response to TBI can lead to apoptotic cell death of oligodendrocytes

Perioperative Neuroscience. DOI: https://doi.org/10.1016/B978-0-323-91003-3.00014-3

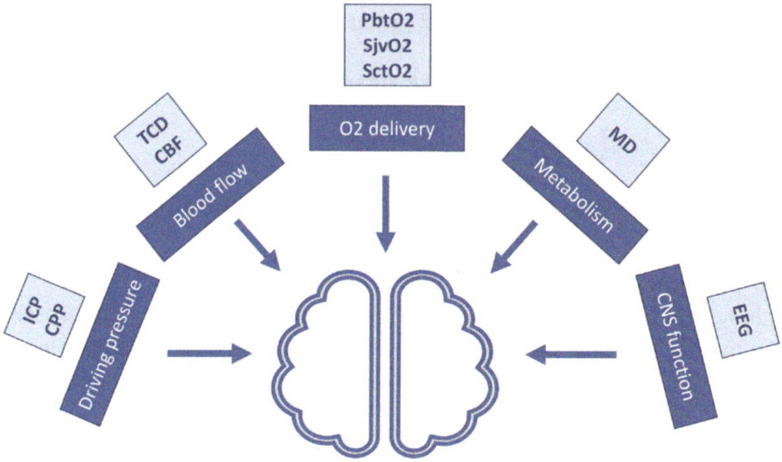

Figure 10.1 Multimodal brain monitoring in traumatic brain injury. *ICP*, Intracranial pressure; *CPP*, cerebral perfusion pressure; *TCD*, trancranial doppler; *CBF*, cerebral blood flow; *PbtO*$_2$, brain oxygen tension pressure; *SjvO*$_2$, jugular venous oxygen saturation; *SctO*$_2$, tissutal cutaneous saturation of oxygen; *MD*, micro dialysis; *EEG*, electroencephalogram; *CNS*, central nervous system.

and neurons, even after a long time from the primary insult.[11] BBB damage and neuronal cells death may be caused by the release of excitatory amino acids such as aspartate and glutamate from presynaptic terminals; this process of excitotoxicity is mediated by calcium passage through the membrane,[12] followed by the production of oxygen reactive species, depolarization of the mitochondrial membrane, and inhibition of adenosine triphosphate synthesis.[13] Calcium can also activate nitric oxide synthetase and induce oxidative, along with the production of reactive oxygen species, damage to proteins and lipids.[14]

Along with molecular mechanisms, intracranial pressure (ICP) is another key element in TBI patients. ICP is a pulsatile pressure, driven by the cardiac cycle and typically defined by three peaks (P1, P2, and P3). The threshold of normal ICP is commonly set at 7–15 mmHg in adults with variation between subjects, and its regulation consists of three major components inside a rigid skull: cerebrospinal fluid (CSF), blood supply, and brain parenchyma. Changes in one of these components can alter the others, thus pressure and volumes are related and maintained within a strict range. Any volume expansion by one intracranial component is normally compensated by changes in the volume of the other components. This buffering capacity commonly starts with the displacement of CSF and venous blood, but when exhausted a rapid increase of ICP is present, until intracranial hypertension (HICP).[15] Values of ICP over 20[16] or 22 mmHg[17] are considered pathological in adults and should be treated with surgery if hematoma is present, or through the escalation therapy (staircase approach),[16] with the goal of cerebral perfusion pressure (CPP) values between 60 and 70 mmHg.[17]

In the case of HICP a "staircase approach" should be pursued.[2,18–20] This escalation therapy starts from basic advisory (tier zero), till the most aggressive treatments (tier one to three). Tier zero includes ICU admission, endotracheal intubation, and mechanical ventilation, head-up 15–30 degrees, analgesia, sedation to prevent ventilator asynchronies, normothermia, end-tidal-CO_2 monitoring, CPP threshold of 60 mmHg, hemoglobin of more than 7 g/dL, normal values of serum sodium, invasive arterial line for continuous pressure monitoring, and oxygen peripheral saturation (SpO_2) $\geq 94\%$.[20] Tier "one" includes analgesia and sedation increase, use of osmotic agents, CSF drainage if external ventricular drainage (EVD) is available, maintenance of partial pressure of carbon dioxide ($PaCO_2$) between 35–38 mmHg, and prophylactic antiseizure drugs. Tier two is represented by mild hypocapnia (32–35 mmHg), neuromuscular paralysis. Tier three includes barbiturates coma, mild hypothermia (35°C–36°C), hyperventilation with a goal of 30–32 mmHg $PaCO_2$, and decompressive craniectomy (DC).[20]

The question/controversy

Despite the availability of new guidelines, it remains an open question on how to manage TBI in the era of novel translational treatments. Indeed, a great variability between centers revealed that high therapy intensity level treatments like metabolic suppression, hypothermia, hyperventilation, and secondary DC are often not proceeded by more conventional lower interventions using an inappropriate escalation strategy while impacting on outcome.[21] Hence, the aim of this chapter is to describe and contextualize conventional and nonconventional pharmacological and nonpharmacological treatments to manage TBI and HICP.

Pharmacological interventions in traumatic brain injury

Sedatives

Propofol is the main agent used for sedation of TBI patients. Sedation and analgesia in TBI patients are normally employed for agitation control, tolerance of endotracheal tube, and to prevent patient–ventilator asynchronies. Midazolam is preferably used when hemodynamic instability is present, although it cannot allow a prompt neurological evaluation.[18] Both these drugs share similar metabolic effects,[22] while the main advantage of using propofol over midazolam is the rapid metabolism to assess a daily neurological wake-up test.[18] Barbiturates use is currently reduced, and they are actually employed only in case of refractory HICP to suppress cerebral electrical activity, metabolic demand, and brainstem reflexes, until reaching burst suppression although it should not be considered the goal of barbiturates.[23]

Other anesthetics and sedatives are currently investigated for the treatment of TBI providing neuroprotection in animal models of TBI.[24] Ketamine provides lower cognitive dysfunction, delirium, and attenuates status epilepticus, and seems to be protective in TBI,[25,26] although its usage alone is not recommended.[18]

Dexmedetomidine showed protective effects on cerebral edema and apoptosis while improving behavioral functions and neurological outcome[27–30] both in preclinical and clinical models.

Similarly, inhaled anesthetics showed promising results.[31] Sevoflurane, xenon, and argon are new interesting topics of research. Xenon improved survival in a mice model of TBI, reducing secondary brain injury.[32] Xenon is also suggested for neurosurgical procedures to protect the brain,[33] while Argon has been tested in both preclinical and clinical models for its surprising neuroprotective characteristics.[34,35]

Osmotic agents

Osmotic agents such as mannitol and hypertonic saline are the drugs most commonly employed for effectively reducing ICP. The superiority of one over the other has not been strongly documented and both increase plasma osmolarity.[36–39] Other characteristics of hypertonic saline are correction of microcirculation and cerebral blood flow and improvement of brain tissue oxygen tension.[40] Moreover, it seems to suppress the production of proinflammatory cytokines in activated microglia.[41]

Hypertonic saline determines a gradient between the vascular compartment and the brain, inducing the passage of free water across the BBB to the bloodstream, with reduction of ICP.[42] Mannitol is a mannose sugar alcohol, which moves free water from brain tissue to the interstitial and vascular compartment; it is not reabsorbed by the renal tubules, thus modifying the osmolarity of glomerular filtrate, inhibiting sodium and chloride reabsorption and stimulating diuresis.[43] Possible side effects of mannitol include dehydration, hypovolemia, and renal tubular epithelial cells injury,[44] whereas hypertonic saline may cause dangerous hypernatremia.[45,46] However, these drugs remain the most adopted in all steps of the "staircase approach" in the case of HICP.[2,18–20]

Mannitol is commonly used at a dose of 0.25–1 g/kg every 4–6 h, whereas hypertonic saline concentrations vary from 3% to 7% and even 23.4%. Their effects persist for 4–6 h, and they also determine hemodilution, increased cardiac output, and blood pressure.[43] Although current guidelines for severe TBI management suggest mannitol (at a dose of 0.25–1 g/kg body weight) over hypertonic saline for HICP control, the debate is still open.[43] Recent studies did not show a superiority of mannitol with respect to hypertonic saline on long-term efficacy and safety after TBI.[43] A useful strategy could be to test both agents with an equimolar bolus, in order to evaluate which approach is more efficacy for each patient.[47]

Continuous infusion of hyperosmolar therapies (whether hypertonic saline or mannitol) is actually not recommended, because of the presence of only limited retrospective data.[48] These data showed that the potential efficacy of continuous infusion over bolus could be related to the patient's osmolarity, but in the absence of more conclusive evidence, hyperosmolar therapies should be administered by bolus.[43]

Analgesics

Opioids are often used in ICU to allow more tolerance to invasive devices and patients' comfort.[49] In the setting of TBI, they can be used for their neuroprotective effects, especially in the early phase of brain trauma. No specific recommendations are available on which agent is preferred over another, but recent evidence confirmed that fentanyl is associated with lower ICP and CPP than sufentanyl and morphine.[49] Opioids generally do not induce effects on ICP, CPP, or mean arterial pressure, except for recent evidence that confirmed that fentanyl is associated with lower ICP and CPP than sufentanyl and morphine.[49] Continuous infusion is associated with a better outcome than bolus.[49]

Antiepileptics

Posttraumatic epilepsy may potentiate secondary brain damage after TBI. Prophylactic antiepileptic drugs have been proposed in the literature to reduce their occurrence, although without strong evidences[50] latest guidelines do not recommend their use as prophylactic agents.[37] The American Academy of Neurology and the Brain Trauma Foundation did not recommend antiepileptic prophylaxis. However, only for the first 7 days after TBI, the use of phenytoin may be considered when the benefits exceed possible complications of treatment. Phenytoin is the drug most reported in the literature as a possible prophylactic agent, which acts by stabilizing the membrane against hyperexcitability while promoting sodium efflux from neurons. Levetiracetam is a newer antiepileptic drug that works by inhibiting burst firing of neurons without acting on neuronal excitability, thus showing promising results in TBI's seizures. Although it has promising efficacy, levetiracetam is not recommended over phenytoin in preventing posttraumatic seizures and toxicity.[2,18,20] Among others, carbamazepine is currently poorly used for prophylaxis and treatment of epilepsy in TBI, as well as phenobarbital which is predominantly used for treating refractory HICP.[18]

Infections prophylaxis

Prophylactic antibiotics are not recommended by the latest guidelines, although antimicrobial-impregnated catheters may be considered to prevent catheter-related infections of EVD.[2] Prophylactic antibiotics for the prevention of other infections are always discouraged. Particularly, the use of povidone-iodine oral care to reduce ventilator-associated pneumonia (VAP) rates may increase the risk of acute respiratory distress syndrome (ARDS). The use of prophylactic antibiotics for the prevention of VAP reduces the rate of infection but not the length of stay and mortality.[51] Uncertain pieces of evidence from an Italian consensus of experts concluded that no evidence is available of antibiotic prophylaxis lasting two or more days in VAP prevention in comatose patients.[52]

Also, early tracheostomy is recommended to reduce the days of mechanical ventilation, but no advantages are present in terms of infections.[2]

Steroids

Corticosteroids are frequently adopted to reduce vasogenic edema and swelling secondary to cerebral neoplasms. Various experimental models have demonstrated their beneficial impact in reducing biochemical and metabolic damages (lipid peroxidation, oxygen radicals).[53] In a clinical setting, two randomized controlled trials (RCTs) demonstrated benefits, while others did not.[54] Thus, current guidelines do not recommend their use in TBI.[2]

Tranexamic acid

Tranexamic acid (TXA) reduces bleeding by inhibiting fibrinolysis, which is the enzymatic process leading to the breakdown of fibrin blood clot.[55] Most severe TBI patients present intracranial bleeding immediately after impact, with possible hematoma expansion occurring within 1–1.5 h of injury[56] and consequent ICP increase causing cerebral herniation and death. TXA has demonstrated bleeding and death reduction in surgery and after traumatic and postpartum hemorrhage.[57] The CRASH-3[55] randomized placebo-controlled trial suggested that timely TXA treatment might reduce intracranial bleeding and early bleeding-related deaths. TXA is administrated at a dose of 1-g bolus within 3 h of injury, followed by an infusion of 1 g over 8 h. TXA half-life is 2 h and it is completely eliminated by the second day.[55] The CRASH-3 trial randomized 9202 patients within 3 h of injury with a GCS score ≤ 12 or intracranial bleeding on CT scan and no significant extracranial bleeding to receive TXA or placebo, observing that TXA administration within 3 h reduces head injury-related death with no evidence of adverse effects or complications.[55]

A recent analysis of CRASH-3 trial studied the effects of TXA on the overall mortality within 24 h of injury and within 28 days, excluding patients with GCS 3 or bilateral unreactive pupil.[56] The study concluded that TXA administration reduces the early deaths in nonmoribund TBI patients but does not modify the risk of death beyond 24 h of injury.[56]

A multicenter cohort study from the Brain Injury: Prehospital Registry of Outcome, Treatments, and Epidemiology of Cerebral Trauma (BRAIN-PROTECT) study in the Netherlands analyzed the outcome of patients with suspected severe TBI in whose administration of TXA was made during prehospital phase. They observed a significant increase in 30-day mortality for patients with severe isolated TBI who received prehospital TXA, concluding that its use is not recommended in the prehospital phase in absence of evidence of extracranial hemorrhage.[58]

Deep vein thrombosis prophylaxis

Deep venous thrombosis prophylaxis should account for the risk of hematoma expansion in case of intracranial hemorrhage. However, pharmacological prophylaxis is indicated in TBI patients, as well as mechanical prophylaxis. Low molecular weight heparins (LMWH) or low-dose unfractioned heparin should be considered in patients with stable TBI.[2] In 2017 a comparative study on 20,417 patients with TBI concluded superiority of LMWH over unfractioned heparin.[59] Using an essay-guided strategy (anti-Xa) over a fixed dose may be promising for effectively reduce the risk of intracranial hemorrhage progression.[60] A RCT confirmed no increased risk of bleeding in the group that received enoxaparin 30 mg bid more than the placebo group.[61] Moreover, a mice model confirmed that edema in the hippocampus and parietal cortex can be reduced by heparin treatment. Also, cognitive impairment was noted in the group treated with enoxaparin.[62] Concerning the timing, few available literatures are present in the clinical scenario. In mice model treated with enoxaparin at 24, 72 or nontreated the rate of further hemorrhage was similar.[63] Similarly, in the clinical setting, starting the prophylaxis 48–72 h post-trauma does not increase the risk of expanding intracranial hemorrhage.[64]

Novel translational drugs

The treatment of cognitive and neuropsychiatric symptoms after TBI is gaining attention. Neurodegeneration can enhance the impairment of network activity and connections of brain areas.[65] Many drugs can alleviate cognitive dysfunction and implement recovery after TBI, but no specific drugs have been approved by the Food and Drug administration in patients with TBI.[66] Methylphenidate, lithium, quetiapine, phenserine tartrate, memantine, and galantamine showed interesting properties, including improvement of cognitive outcomes (methylphenidate),[67] increase of synaptic density, and modulation of dopamine neurotransmission (quetiapine),[68] alleviation of oxidative stress, glutamate excitotoxicity, and neuroinflammation (phenserine tartrate),[69] reduction of astrogliosis, cell death, and functional deficit in in vitro and in vivo models of TBI (memantine and galantamine).[70,71]

Among other drugs, methylene blue should be mentioned for its promising efficacy. Acute administration of methylene blue improved neuroinflammation and behavioral functions,[72] and reduced the lesion volume on quantitative magnetic resonance.[73] Moreover, methylene blue seems to be able of reversing mitochondrial damage, reducing neuronal apoptosis, and improving BBB integrity after TBI.[74] Unfortunately, no clinical data in patients are available. In rats, statins reduced neuronal apoptosis of microglia, tumor necrosis factor expression, while resulting in protective cognitive dysfunction.[75,76] In a clinical setting, less evidence is available but in TBI, statins resulted associated with good clinical outcome.[77]

Protective brain effects of erythropoietin have been confirmed by several preclinical models. Erythropoietin prevented loss of tight junctions and BBB dysfunction attenuated vasogenic edema and modulated mitochondrial activity.[78,79]

Other natural and synthetic compounds have been proposed for the treatment and management of TBI. However, they are still currently employed in preclinical setting only in the majority of cases.[80]

Nonpharmacological interventions in traumatic brain injury

Ventilatory management

A recent consensus on how to ventilate a patient with acute brain injury has been published.[81] Despite the lack of consistent literature regarding this topic, several recommendations have been carried out. A threshold of arterial partial pressure of oxygen (PaO_2) 80–120 mmHg and arterial $PaCO_2$ 35–45 mmHg should be pursued both in patients with or without ARDS and both in case of HICP or not. In the case of HICP refractory to conventional "first tiers" therapies, hyperventilation should be considered.[18,20,81] Lung protective mechanical ventilation should be used in the case of non-HICP. The same level of positive end expiratory pressure (PEEP) should be used as in patients without brain injury if PEEP is insensitive and without ARDS.[81]

Moreover, in the case of HICP, hyperventilation should be considered as part of the escalation therapy as above mentioned. In fact, during tier "one" $PaCO_2$ should be ranged between 35–38 mmHg, during tier two mild hypocapnia (32–35 mmHg) can be achieved, whereas in tier three hyperventilation with a goal of 30–32 mmHg $PaCO_2$ may be necessary before more aggressive treatments.[20]

CSF drainage

CSF drainage revests a pivotal role in TBI management, especially during HICP.[18,43] In fact, continuous CSF drainage may be helpful if EVD is available, although there is no strong evidence of its efficacy on long-term intracranial hypertension control.[16,43] Intraventricular ICP monitoring devices are catheters with a transducer (fiberoptic strain gauge or pneumatic sensor) placed into the cerebral ventricles, which are connected to an external pressure monitoring system.[16,82] This type of device consists of an EVD, which allows CSF drainage through the catheter in case of intracranial hypertension, contributing to ICP control.[16,18] Intraventricular ICP is the oldest technique described by Guillaume and Lundberg,[83,84] and still remains the more reliable system to detect ICP values and intracranial compliance.[82,85] Misplacement of the device, dislocation, kinks, obstruction from blood and tissue debris, hemorrhage, and infection are the possible complication related to intraventricular ICP monitoring.[43] Dislocation, kinks, and obstruction are reported in 4%, 5%–25%,[43] whereas hemorrhage ranges from 0.7% to 0.61% considering only the symptomatic hemorrhage,[86,87] and 2.5% considering all the bleedings after EVD placement.[88] Infection ranges from 1 to 27%, and it is related to the duration of maintenance of the device; antibiotic and silver-impregnated catheters and the short use of EVD (less than 5 days) have shown a reduction of the infection rate.[18,43] In conclusion, intraventricular ICP monitoring is a high reliable methodic for the management of TBI patients, representing the gold standard for intracranial compliance evaluation.[85] However, it is associated with potential complications which lead to the need for careful management of the device. Moreover, the intraventricular catheter could be difficult to position in a younger patient due to the smaller ventricle volume, and in some TBI patients with HICP in which the compensatory mechanisms determine the collapse of the ventricular system.[85] Although intraventricular devices are usually considered the "gold standard" for ICP monitoring and management, studies comparing this type of device with intraparenchymal transducers have shown no clear results.[89–92]

Decompressive craniectomy

The removal of a portion of the skull in order to reduce HICP is called DC, which can be primary or secondary. Primary DC refers to the removal of a bone flap after evacuating an intracranial mass lesion. Secondary DC consists of the removal of a portion of the skull after HICP.[93] Both techniques have shown a higher efficacy in reducing and treating HICP. Besides, the optimal timing and indications for DC have been largely investigated.[93] The latest guidelines for the management of primary DC have suggested considering leaving the bone flap out based upon radiological and clinical examinations, after evacuating a subdural hematoma if the brain is bulging. In case of evacuation of epidural hematoma, the bone flap should be replaced. In primary DC, the bone flap should be as larger as 12 cm × 15 cm.[93] In every case of bone replacement, an ICP probe should be positioned, while in case of nonpositioning a CT scan should be used for monitoring progression.[93] For intermediate or doubt categories, the choice is of surgeon's judgment. In the case of secondary DC, ICP monitoring is considered an essential tool for guiding therapy, since secondary DC reduces ICP, increases CPP, and may decrease mortality.[93]

DC compared with medical management has been investigated by two multicenter RCTs, which tried to provide guidance to clarify the timing and indications: Decompressive Craniectomy in Patients with Severe Traumatic brain Injury (DECRA) and Trial of Decompressive Craniectomy for Traumatic Intracranial Hypertension (RESCUEicp).[94,95] The DECRA study found similar mortality between the medical and surgical cohorts, with a higher unfavorable neurologic outcome among the surgical group.[95] On the contrary, the RESCUEicp trial showed lower mortality for DC, higher rates of vegetative state, as well as lower and upper severe disability at 6 months, with respect to medical therapy.[94] The fundamental difference between the two trials was the inclusion criteria, which led the DECRA trial to investigate the effects of DC for *early* HICP, whereas led RESCUEicp to analyze the effects of DC for *late* HICP.[96] In fact, DECRA inclusion criteria was HICP ($>$20 mmHg) for 15 min over a 1-h period although the tier 1 therapies within the first 72 h after trauma, while RESCUEicp included cases with HICP ($>$25 mmHg) for 1–12 h despite the tiers 1 and 2 therapies within the first 10 days after TBI.[94,95] A recent update on DC by Brain Trauma Foundation based on RESCUEicp and DECRA developed Level IIA recommendations on DC, suggesting that secondary DC is recommended in case of late refractory HICP for outcome improvement, whereas it is not suggested to treat early refractory HICP.[96]

Translational evidences

Therapeutic hypothermia showed promising results in animal studies.[97] However, clinical studies did not confirm the same conclusions. In some cases, therapeutic hypothermia played a negative role in mortality and functional outcome.[98,99] Several systematic reviews confirmed the utility of therapeutic hypothermia.[100–102] A recent metaanalysis reported benefits in terms of mortality, but it was based on poor-quality studies.[102] Controversy, another metaanalysis confirmed different results.[99] Moreover, the most recent metaanalysis in a total of 2796 patients concluded lower mortality within 24 h. Moreover, the risk of unfavorable outcomes was reduced after therapeutic hypothermia; and a higher risk of pneumonia was found.[103]

Two major RCTs on hypothermia after severe TBI have investigated the effectiveness of hypothermia (32°C−35°C) to reduce ICP and morbidity and mortality after TBI: the Eurotherm3235 RCT and the POLAR Randomized Clinical Trial.[104,105] The Eurotherm trial analyzed patient receiving hypothermia (32°C−35°C) added to standard care and observed that therapeutic hypothermia successfully controlled ICP but determined higher mortality and worse functional outcome at 6 months.[105]

The POLAR trial compared the use of early prophylactic hypothermia (32°C−35°C) with normothermia, showing no improvement of neurologic outcome and mortality at 6 months, thus not supporting its use for patients with severe TBI.[104]

Consensus statement

During the management of TBI, a multimodal neuromonitoring strategy is always suggested when deemed available, thus helping the clinician to undergo appropriate management. Despite the existence of new guidelines, the appropriate management of TBI and HICP is still under investigation, especially regarding nonconventional pharmacological treatments.

Conclusion

TBI treatment and management are continuously evolving. Current guidelines do not include the use of novel pharmacological and nonpharmacological agents yet, but experimental and clinical data are showing really promising results to improve and treat TBI in the next future with novel translational strategies.

Conflict of interest

All authors declare no conflict of interest. This chapter did not receive any specific grant from funding agencies in the public, commercial, or not-for-profit sectors.

Authors' contribution

D.B., conceptualization, writing—review and editing; C.R., P.P. revising it critically for important intellectual content, and final approval of the version submitted.

References

1. Dewan MC, Rattani A, Gupta S, et al. Estimating the global incidence of traumatic brain injury. *J Neurosurg*. 2019;130:1080−1097.

2. Carney N, Totten AM, O'reilly C, et al. Guidelines for the management of severe traumatic brain injury, fourth edition. *Neurosurgery*. 2017;80:6–15.
3. Gruenbaum SE, Zlotnik A, Gruenbaum BF, Hersey D, Bilotta F. Pharmacologic neuroprotection for functional outcomes after traumatic brain injury: a systematic review of the clinical literature. *CNS Drugs*. 2016;30:791–806.
4. Mckee AC, Daneshvar DH. The neuropathology of traumatic brain injury. *Handb Clin Neurol*. 2015;127:45–66.
5. Corps KN, Roth TL, McGavern DB. Inflammation and neuroprotection in traumatic brain injury. *JAMA Neurol*. 2015;72:355.
6. Kwiecien JM, Dabrowski W, Dąbrowska-Bouta B, et al. Prolonged inflammation leads to ongoing damage after spinal cord injury. *PLoS One*. 2020;15:e0226584.
7. Karimi-Abdolrezaee S, Billakanti R. Reactive astrogliosis after spinal cord injury-beneficial and detrimental effects. *Mol Neurobiol*. 2012;46:251–264.
8. Lazaridis C, Rusin CG, Robertson CS. Secondary brain injury: predicting and preventing insults. *Neuropharmacology*. 2019;145:145–152.
9. Pisoschi AM, Pop A. The role of antioxidants in the chemistry of oxidative stress: a review. *Eur J Medicinal Chem*. 2015;97:55–74.
10. Büki A, Povlishock JT. All roads lead to disconnection?—traumatic axonal injury revisited. *Acta Neurochir (Wien)*. 2006;148:181–193. discussion 193–4.
11. Mu J, Li M, Wang T, et al. Myelin damage in diffuse axonal injury. *Front Neurosci*. 2019;13.
12. Tehse J, Taghibiglou C. The overlooked aspect of excitotoxicity: glutamate-independent excitotoxicity in traumatic brain injuries. *Eur J Neurosci*. 2019;49:1157–1170.
13. Singh IN, Sullivan PG, Deng Y, Mbye LH, Hall ED. Time course of post-traumatic mitochondrial oxidative damage and dysfunction in a mouse model of focal traumatic brain injury: implications for neuroprotective therapy. *J Cereb Blood Flow Metab*. 2006;26:1407–1418.
14. Hall ED, Detloff MR, Johnson K, Kupina NC. Peroxynitrite-mediated protein nitration and lipid peroxidation in a mouse model of traumatic brain injury. *J Neurotrauma*. 2004;21:9–20.
15. Canac N, Jalaleddini K, Thorpe SG, Thibeault CM, Hamilton RB. Review: pathophysiology of intracranial hypertension and noninvasive intracranial pressure monitoring. *Fluids Barriers CNS*. 2020;17:40.
16. Stocchetti N, Maas AIR. Traumatic intracranial hypertension. *N Engl J Med*. 2014;370:2121–2130.
17. Carney N, Totten AM, O'Reilly C, et al. Guidelines for the management of severe traumatic brain injury, fourth edition. *Neurosurgery*. 2017;80:6–15.
18. Battaglini D, Anania P, Rocco PRM, et al. Escalate and de-escalate therapies for intracranial pressure control in traumatic brain injury. *Front Neurol*. 2020;11:564751.
19. Chesnut RM, Temkin N, Videtta W, et al. Consensus-based management protocol (CREVICE Protocol) for the treatment of severe traumatic brain injury based on imaging and clinical examination for use when intracranial pressure monitoring is not employed. *J Neurotrauma*. 2020;37:1291–1299.
20. Chesnut R, Aguilera S, Buki A, et al. A management algorithm for adult patients with both brain oxygen and intracranial pressure monitoring: the Seattle International Severe Traumatic Brain Injury Consensus Conference (SIBICC). *Intensive Care Med*. 2020;46:919–929.
21. Huijben JA, Dixit A, Stocchetti N, et al. Use and impact of high intensity treatments in patients with traumatic brain injury across Europe: a CENTER-TBI analysis. *Crit Care*. 2021;25:78.

22. Tanguy M, Seguin P, Laviolle B, Bleichner JP, Morandi X, Malledant Y. Cerebral microdialysis effects of propofol vs midazolam in severe traumatic brain injury. *J Neurotrauma*. 2012;29:1105–1110.
23. Stocchetti N, Zanaboni C, Colombo A, et al. Refractory intracranial hypertension and "second-tier" therapies in traumatic brain injury. *Intensive Care Med*. 2008;34: 461–467.
24. Archer DP, McCann SK, Walker AM, et al. Neuroprotection by anaesthetics in rodent models of traumatic brain injury: a systematic review and network *meta*-analysis. *Br J Anaesth*. 2018;121:1272–1281.
25. Bell JD. In vogue: ketamine for neuroprotection in acute neurologic injury. *Anesthesia Analgesia*. 2017;124:1237–1243.
26. Chang LC, Raty SR, Ortiz J, Bailard NS, Mathew SJ. The Emerging use of ketamine for anesthesia and sedation in traumatic brain injuries. *CNS Neurosci Ther*. 2013;19: 390–395.
27. Li F, Wang X, Deng Z, Zhang X, Gao P, Liu H. Dexmedetomidine reduces oxidative stress and provides neuroprotection in a model of traumatic brain injury via the PGC-1α signaling pathway. *Neuropeptides*. 2018;72:58–64.
28. Zhang MH, Zhou XM, Cui JZ, Wang KJ, Feng Y, Zhang HA. Neuroprotective effects of dexmedetomidine on traumatic brain injury: involvement of neuronal apoptosis and HSP70 expression. *Mol Med Rep*. 2018;17:8079–8086.
29. Shen M, Wang S, Wen X, et al. Dexmedetomidine exerts neuroprotective effect via the activation of the PI3K/Akt/mTOR signaling pathway in rats with traumatic brain injury. *Biomed Pharmacother*. 2017;95:885–893.
30. Humble SS, Wilson LD, Leath TC, et al. ICU sedation with dexmedetomidine after severe traumatic brain injury. *Brain Inj*. 2016;30:1266–1270.
31. Wan Hassan WMN, Mohd Nasir Y, Mohamad Zaini RH, Wan Muhd Shukeri WF. Target-controlled infusion propofol vs sevoflurane anaesthesia for emergency traumatic brain surgery: comparison of the outcomes. *Malays J Med Sci*. 2017;24:73–82.
32. Campos-Pires R, Hirnet T, Valeo F, et al. Xenon improves long-term cognitive function, reduces neuronal loss and chronic neuroinflammation, and improves survival after traumatic brain injury in mice. *Br J Anaesth*. 2019;123:60–73.
33. Rylova A, Maze M. Protecting the brain with xenon anesthesia for neurosurgical procedures. *J Neurosurg Anesthesiol*. 2019;31:18–29.
34. Koziakova M, Harris K, Edge CJ, Franks NP, White IL, Dickinson R. Noble gas neuroprotection: xenon and argon protect against hypoxic–ischaemic injury in rat hippocampus in vitro via distinct mechanisms. *Br J Anaesth*. 2019;123:601–609.
35. Höllig A, Schug A, Fahlenkamp A, Rossaint R, Coburn M. Argon organo-protective network (AON). Argon: systematic review on neuro- and organoprotective properties of an "inert" gas. *Int J Mol Sci*. 2014;15:18175–18196.
36. Raslan A, Bhardwaj A. Medical management of cerebral edema. *Neurosurg Focus*. 2007;22:E12.
37. Hawryluk GWJ, Aguilera S, Buki A, et al. A management algorithm for patients with intracranial pressure monitoring: the Seattle International Severe Traumatic Brain Injury Consensus Conference (SIBICC). *Intensive Care Med*. 2019;1783–1794.
38. Gu J, Huang H, Huang Y, Sun H, Xu H. Hypertonic saline or mannitol for treating elevated intracranial pressure in traumatic brain injury: a *meta*-analysis of randomized controlled trials. *Neurosurg Rev*. 2019;42:499–509.
39. Cook AM, Morgan Jones G, Hawryluk GWJ, et al. Guidelines for the acute treatment of cerebral edema in neurocritical care patients. *Neurocrit Care*. 2020;32:647–666.

40. Rockswold GL, Solid CA, Paredes-Andrade E, Rockswold SB, Jancik JT, Quickel RR. Hypertonic saline and its effect on intracranial pressure, cerebral perfusion pressure, and brain tissue oxygen. *Neurosurgery*. 2009;65:1035–1041.
41. Zeng WX, Han YL, Zhu GF, et al. Hypertonic saline attenuates expression of notch signaling and proinflammatory mediators in activated microglia in experimentally induced cerebral ischemia and hypoxic BV-2 microglia. *BMC Neurosci*. 2017;18:32.
42. Himmelseher S, Durieux ME. Revising a dogma: ketamine for patients with neurological injury? *Anesth Analg*. 2005;101:524–534.
43. Anania P, Battaglini D, Miller JP, et al. Escalation therapy in severe traumatic brain injury: how long is intracranial pressure monitoring necessary? *Neurosurg Rev*. 2020; 1–9.
44. Robba C, Banzato E, Rebora P, et al. Acute kidney injury in traumatic brain injury patients: results from the collaborative European neurotrauma effectiveness research in traumatic brain injury study. *Crit Care Med*. 2021;49:112–126.
45. Nomani AZ, Nabi Z, Rashid H, et al. Osmotic nephrosis with mannitol: review article. *Ren Fail*. 2014;36:1169–1176.
46. Shi J, Qian J, Li H, Luo H, Luo W, Lin Z. Renal tubular epithelial cells injury induced by mannitol and its potential mechanism. *Ren Fail*. 2018;40:85–91.
47. Robba C, Citerio G. How I manage intracranial hypertension. *Crit Care*. 2019;23:243.
48. Ropper AH. Hyperosmolar therapy for raised intracranial pressure. *N Engl J Med*. 2012;367:746–752.
49. Wiener J, McIntyre A, Janzen S, Mirkowski M, MacKenzie HM, Teasell R. Opioids and cerebral physiology in the acute management of traumatic brain injury: a systematic review. *Brain Inj*. 2019;33:559–566.
50. Chartrain AG, Yaeger K, Feng R, et al. Antiepileptics for post-traumatic seizure prophylaxis after traumatic brain injury. *Curr Pharm Des*. 2018;23:6428–6441.
51. Reizine F, Asehnoune K, Roquilly A, et al. Effects of antibiotic prophylaxis on ventilator-associated pneumonia in severe traumatic brain injury. A post hoc analysis of two trials. *J Crit Care*. 2019;50:221–226.
52. Poole D, Chieregato A, Langer M, et al. Systematic review of the literature and evidence-based recommendations for antibiotic prophylaxis in trauma: results from an Italian consensus of experts. *PLoS One*. 2014;9:e113676.
53. Hoshide R, Cheung V, Marshall L, Kasper E, Chen C. Do corticosteroids play a role in the management of traumatic brain injury? *Surg Neurol Int*. 2016;7:84.
54. Roberts I, Yates D, Sandercock P, et al. Effect of intravenous corticosteroids on death within 14 days in 10 008 adults with clinically significant head injury (MRC CRASH trial): randomised placebo-controlled trial. *Lancet*. 2004;364:1321–1328.
55. Collaborators C-3 trial. Effects of tranexamic acid on death, disability, vascular occlusive events and other morbidities in patients with acute traumatic brain injury (CRASH-3): a randomised, placebo-controlled trial. *Lancet*. 2019;394:1713–1723.
56. Brenner A, Belli A, Chaudhri R, et al. Understanding the neuroprotective effect of tranexamic acid: an exploratory analysis of the CRASH-3 randomised trial. *Crit Care*. 2020;24:560.
57. CRASH-2 trial collaborators. Effects of tranexamic acid on death, vascular occlusive events, and blood transfusion in trauma patients with significant haemorrhage (CRASH-2): a randomised, placebo-controlled trial. *Lancet*. 2010;376:23–32.
58. Bossers SM, Loer SA, Bloemers FW, et al. Association between prehospital tranexamic acid administration and outcomes of severe traumatic brain injury. *JAMA Neurol*. 2021;78:338.

59. Benjamin E, Recinos G, Aiolfi A, Inaba K, Demetriades D. Pharmacological thromboembolic prophylaxis in traumatic brain injuries. *Ann Surg*. 2017;266:463–469.
60. Rodier SG, Kim M, Moore S, et al. Early anti-xa assay-guided low molecular weight heparin chemoprophylaxis is safe in adult patients with acute traumatic brain injury. *Am Surg*. 2020;86:369–376.
61. Phelan HA, Wolf SE, Norwood SH, et al. A randomized, double-blinded, placebo-controlled pilot trial of anticoagulation in low-risk traumatic brain injury. *J Trauma Acute Care Surg*. 2012;73:1434–1441.
62. Wahl F, Grosjean-Piot O, Bareyre F, Uzan A, Stutzmann J-M. Enoxaparin reduces brain edema, cerebral lesions, and improves motor and cognitive impairments induced by a traumatic brain injury in rats. *J Neurotrauma*. 2000;17:1055–1065.
63. Dhir T, Weiss E, Wolanin K, et al. The impact of enoxaparin administration in relationship to hemorrhage in mild traumatic brain injury. *Injury*.. 2018;49:2174–2177.
64. Dudley RR, Aziz I, Bonnici A, et al. Early venous thromboembolic event prophylaxis in traumatic brain injury with low-molecular-weight heparin: risks and benefits. *J Neurotrauma*. 2010;27:2165–2172.
65. Gabrieli D, Schumm SN, Vigilante NF, Parvesse B, Meaney DF. Neurodegeneration exposes firing rate dependent effects on oscillation dynamics in computational neural networks. *PLoS One*. 2020;15:e0234749.
66. Rabinowitz AR, Watanabe TK. Pharmacotherapy for treatment of cognitive and neuropsychiatric symptoms after mTBI. *J Head Trauma Rehabil*. 2020;35:76–83.
67. Chien YJ, Chien YC, Liu CT, Wu HC, Chang CY, Wu MY. Effects of methylphenidate on cognitive function in adults with traumatic brain injury: a *meta*-analysis. *Brain Sci*. 2019;9:E291.
68. De Sarno P, Li X, Jope RS. Regulation of Akt and glycogen synthase kinase-3β phosphorylation by sodium valproate and lithium. *Neuropharmacology*. 2002;43:1158–1164.
69. Lilja AM, Luo Y, Yu QS, et al. Neurotrophic and neuroprotective actions of (−)- and (+)-phenserine, candidate drugs for Alzheimer's disease. *PLoS One*. 2013;8:e54887.
70. Effgen GB, Morrison B. Memantine reduced cell death, astrogliosis, and functional deficits in an in vitro model of repetitive mild traumatic brain injury. *J Neurotrauma*. 2017;34:934–942.
71. Day NL, Carle MS, Floyd CL. Post-injury administration of a combination of memantine and 17β-estradiol is protective in a rat model of traumatic brain injury. *Neurochem Int*. 2017;111:57–68.
72. Fenn AM, Skendelas JP, Moussa DN, et al. Methylene blue attenuates traumatic brain injury-associated neuroinflammation and acute depressive-like behavior in mice. *J Neurotrauma*. 2015;32:127–138.
73. Watts LT, Long JA, Boggs RC, et al. Delayed methylene blue improves lesion volume, multi-parametric quantitative magnetic resonance imaging measurements, and behavioral outcome after traumatic brain injury. *J Neurotrauma*. 2016;33:194–202.
74. Shen J, Xin W, Li Q, Gao Y, Yuan L, Zhang J. Methylene blue reduces neuronal apoptosis and improves blood-brain barrier integrity after traumatic brain injury. *Front Neurol*. 2019;10.
75. Lim SW, Shiue YL, Liao JC, et al. Simvastatin therapy in the acute stage of traumatic brain injury attenuates brain trauma-induced depression-like behavior in rats by reducing neuroinflammation in the hippocampus. *Neurocrit Care*. 2017;26:122–132.
76. Mountney A, Boutté AM, Gilsdorf J, Lu XC, Tortella FC, Shear DA. Intravenous administration of simvastatin improves cognitive outcome following severe traumatic brain injury in rats. *J Neurotrauma*. 2016;33:1492–1500.

77. Lokhandwala A, Hanna K, Gries L, et al. Preinjury statins are associated with improved survival in patients with traumatic brain injury. *J Surg Res*. 2020;245:367–372.
78. Blixt J, Gunnarson E, Wanecek M. Erythropoietin attenuates the brain edema response after experimental traumatic brain injury. *J Neurotrauma*. 2018;35:671–680.
79. Millet A, Bouzat P, Trouve-Buisson T, et al. Erythropoietin and its derivates modulate mitochondrial dysfunction after diffuse traumatic brain injury. *J Neurotrauma*. 2016; 33:1625–1633.
80. Battaglini D, Siwicka-Gieroba D, Rocco PR, et al. Novel synthetic and natural therapies for traumatic brain injury. *Curr Neuropharmacol*. 2021;19. Ahead of print.
81. Robba C, Poole D, Monet M, et al. Mechanical ventilation in patients with acute brain injury: recommendations of the European Society of Intensive Care Medicine consensus. *Intensive Care Med*. 2020;46:2397–2410.
82. Bales JW, Bo now RH, Buckley RT, Barber J, Temkin N, Chesnut RM. Primary external ventricular drainage catheter vs intraparenchymal ICP monitoring: outcome analysis. *Neurocrit Care*. 2019;31:11–21.
83. Guillaume J, Janny P. Continuous intracranial manometry; importance of the method and first results. *Rev Neurol*. 1951;84:131–142.
84. Lundberg N. Continuous recording and control of ventricular fluid pressure in neurosurgical practice. *Acta Psychiatr Scand Suppl*. 1960;36(149):1–193.
85. Nag DS, Sahu S, Swain A, Kant S. Intracranial pressure monitoring: gold standard and recent innovations. *World J Clin Cases*. 2019;7:1535–1553.
86. Bauer DF, Razdan SN, Bartolucci AA, Markert JM. *Meta*-analysis of hemorrhagic complications from ventriculostomy placement by neurosurgeons. *Neurosurgery*. 2011;69: 255–260.
87. Binz DD, Toussaint LG, Friedman JA. Hemorrhagic complications of ventriculostomy placement: a *meta*-analysis. *Neurocrit Care*. 2009;10:253.
88. Gelabert-González M, Ginesta-Galan V, Sernamito-García R, Allut AG, Bandin-Diéguez J, Rumbo RM. The Camino intracranial pressure device in clinical practice. Assessment in a 1000 cases. *Acta Neurochir (Wien)*. 2006;148:435–441.
89. Aiolfi A, Khor D, Cho J, Benjamin E, Inaba K, Demetriades D. Intracranial pressure monitoring in severe blunt head trauma: does the type of monitoring device matter? *J Neurosurg*. 2018;128:828–833.
90. Liu H, Wang W, Cheng F, et al. External ventricular drains vs intraparenchymal intracranial pressure monitors in traumatic brain injury: a prospective observational study. *World Neurosurg*. 2015;83:794–800.
91. Kasotakis G, Michailidou M, Bramos A, et al. Intraparenchymal vs extracranial ventricular drain intracranial pressure monitors in traumatic brain injury: less is more? *J Am Coll Surg*. 2012;214:950–957.
92. Volovici V, Huijben JA, Ercole A, et al. Ventricular drainage catheters vs intracranial parenchymal catheters for intracranial pressure monitoring-based management of traumatic brain injury: a systematic review and *meta*-analysis. *J Neurotrauma*. 2019; 36:988–995.
93. Hutchinson PJ, Kolias AG, Tajsic T, et al. Consensus statement from the international consensus meeting on the role of decompressive craniectomy in the management of traumatic brain injury. *Acta Neurochir (Wien)*. 2019;161:1261–1274.
94. Hutchinson PJ, Kolias AG, Timofeev IS, et al. Trial of decompressive craniectomy for traumatic intracranial hypertension. *N Engl J Med*. 2016;375:1119–1130.
95. Cooper DJ, Rosenfeld JV, Murray L, et al. Decompressive craniectomy in diffuse traumatic brain injury. *N Engl J Med*. 2011;364:1493–1502.

96. Hawryluk GWJ, Rubiano AM, Totten AM, et al. Guidelines for the management of severe traumatic brain injury: 2020 update of the decompressive craniectomy recommendations. *Neurosurgery*. 2020;87:427–434.
97. Dietrich WD, Bramlett HM. The evidence for hypothermia as a neuroprotectant in traumatic brain injury. *Neurotherapeutics*. 2010;7:43–50.
98. Andrews PJD, Sinclair HL, Rodriguez A, et al. Hypothermia for intracranial hypertension after traumatic brain injury. *N Engl J Med*. 2015;373:2403–2412.
99. Watson HI, Shepherd AA, Rhodes JKJ, Andrews PJD. Revisited: a systematic review of therapeutic hypothermia for adult patients following traumatic brain injury. *Crit Care Med*. 2018;46:972–979.
100. Crossley S, Reid J, McLatchie R, et al. A systematic review of therapeutic hypothermia for adult patients following traumatic brain injury. *Crit Care*. 2014;18:R75.
101. Zhu Y, Yin H, Zhang R, Ye X, Wei J. Therapeutic hypothermia vs normothermia in adult patients with traumatic brain injury: a *meta*-analysis. *Springerplus*. 2016;5:801.
102. Crompton EM, Lubomirova I, Cotlarciuc I, Han TS, Sharma SD, Sharma P. *Meta*-analysis of therapeutic hypothermia for traumatic brain injury in adult and pediatric patients. *Crit Care Med*. 2017;45:575–583.
103. Chen H, Wu F, Yang P, Shao J, Chen Q, Zheng R. A *meta*-analysis of the effects of therapeutic hypothermia in adult patients with traumatic brain injury. *Crit Care*. 2019;23:396.
104. Cooper DJ, Nichol AD, Bailey M, et al. Effect of early sustained prophylactic hypothermia on neurologic outcomes among patients with severe traumatic brain injury. *JAMA*. 2018;320:2211.
105. Andrews PJ, Sinclair HL, Rodríguez A, et al. Therapeutic hypothermia to reduce intracranial pressure after traumatic brain injury: the Eurotherm3235 RCT. *Health Technol Assess (Rockv)*. 2018;22:1–134.

Movement disorders and translational research

11

Supriyo Choudhury[1], *Sakhi Bhansali*[1], *Mona Tiwari*[2] *and Hrishikesh Kumar*[1]
[1]Department of Neurology, Institute of Neurosciences Kolkata, Kolkata, India,
[2]Department of Neuro-Radiology, Institute of Neurosciences Kolkata, Kolkata, India

Translational medicine is defined by the European Society for Translational Medicine as "an interdisciplinary branch of the biomedical field supported by three main pillars: bench side, bedside, and community".[1] It collates basic science and clinical research to generate practical and tangible solutions to real world health issues. This "bench-to-bedside" enterprise of harnessing knowledge helps clinicians and scientists to understand the disease process better, leading to more suitable and diverse diagnostic and therapeutic options.[2] Additionally, "bedside-to-bench" is a complimentary aspect of translational medicine where interesting clinical observations are explained in laboratory using cutting edge research tools.[1]

Movement disorder (MD) is a unique field wherein specialists rely on clinical phenomenology to form a differential, and an accurate clinical diagnosis is usually made after taking into account the history and neurological examination. Translational research in MDs includes but is not limited to in silico (computer simulations), in vitro (on cell line), and in vivo (animal model) research.[3–6] It also includes studies on human motor control to understand physiology and clinical trials to assess interventions in patients with MD.[7] This chapter encapsulates all these domains and identifies the potential future directions of translational research based on challenges faced in clinical practice. Our aim is to give a bird's eye view of translational research in MD to a beginner. A translational researcher often works as a bridge between a clinician and a basic scientist and should have an overall understanding on both these aspects. Therefore, in addition to recent scientific development in molecular biology, this chapter also incorporates the clinical practices routinely followed in a MD clinic. We have divided this chapter into 2 main sections: the first section deals with- etiology, diagnosis and monitoring, and the second section deals with-prevention and treatment. Each section has several subsections, which have been segmented into Parkinson's disease (PD) and non-PD MD, wherever possible.

Etiology, diagnosis, and monitoring

Biomarkers

The diagnosis of PD mainly relies on the UK Parkinson's Disease Society Brain Bank criteria.[8] In majority of the cases, a clinical diagnosis is accurate, but it could

Perioperative Neuroscience. DOI: https://doi.org/10.1016/B978-0-323-91003-3.00005-2

be challenging at early stages of the disease where close mimickers like essential tremor, drug induced parkinsonism, atypical parkinsonism etc. can be confused with PD. Hence, in the event of a diagnostic dilemma, although impractical, a brain biopsy looking for Lewy bodies is the only confirmatory diagnostic test for PD.[9] To overcome this limitation, efforts to find noninvasive or minimally invasive ways to detect PD with accuracy are underway and this field is growing rapidly. Currently, biochemical, radiological, genetic, and electrophysiological tools are being studied for early detection of PD.[10–13] Kinematic studies of walking pattern (gait) have also been found to be useful.[14] These areas form a major thrust of translational research in PD and other MDs. The goal is to develop one or more reliable and clinically useful biomarkers which aid in diagnosis and monitoring of patients, alongside accelerating the process of drug development. A sensitive biomarker is often the key to success of various clinical trials.[15] Unfortunately, till date, such sensitive biomarkers are elusive in the field of MD.

Huntington's disease (HD) is one of the rare neurodegenerative conditions for which predictive genetic testing is available for individuals with a known family history. This diagnostic biomarker helps identify "pre-manifest" individuals, who are uniquely poised to potentially benefit from neuroprotective therapy, which would delay (or even prevent) the development of the disease's clinical manifestations and functional disability.[16] Even though no disease modifying therapies are currently available, several are in the pipeline.[17] Developing/discovering sensitive biomarkers will help us understand whether these patients respond to future therapies in a predictable and objective manner. Wet biomarkers that reflect the pathobiology of HD are logical candidates being explored. Current research on pathologically relevant molecules such as misfolded huntingtin protein aggregates, in combination with other quantifiable biomarkers looks promising.[16]

Biochemistry

PD is a multiorgan proteinopathy where an abnormal protein named alpha-synuclein (among other proteins) is deposited inside the neuron to form Lewy body and Lewy Neurites throughout Central Nervous System (CNS).[18] The soluble protein often trickles to CSF and blood and is found in other tissues like skin, salivary gland, and gastrointestinal tract.[19] Alpha-synuclein is a house-keeping protein, found in majority of healthy cells but its modified forms like phosphorylated and nitrated compounds might be associated with neurodegeneration.[20] It is still unclear whether the concentration of these proteins in peripheral body fluids and CSF can distinguish PD from matched healthy individual.[21] Studies found that the level of total alpha synuclein is low in body fluids of PD patients compared to healthy.[20] Whereas the modified or oligomeric forms are raised in the CSF or serum of PD patients.[22] The reason of this apparent contradiction is still elusive to a large extent, but researchers speculate that low alpha-synuclein could be due to its sequestration in the intracellular inclusions and a raised oligomeric protein is due to increased production/reduced intracellular trafficking of pathogenic misfolded proteins. Notably, some studies failed to observe any significant difference in these serum-biomarkers between PD and healthy people. It is quite possible that the transport of this protein is incomplete from CNS to blood and hence the difference was not appreciated.[23]

Alpha-Synuclein and its oligomers are potential predictors of disease phenotype in PD. Joseph Jankovic (1990) proposed two broad phenotypical subtypes of PD called tremor dominant (TD) and postural instability and gait dominant (PIGD)[24]. As the name suggest, the phenotypes describe the predominant presentation of the patients. In PIGD variant the primary presentations are problem in balance and walking and it is likely to progress faster than the other variant which mainly present with resting tremor.[24] Reports show, the phosphorylated alpha-synuclein are significantly higher in PIGD group compared to TD.[25] Hence, specific biomarkers of these phenotypes are not just useful to predict the progress but also to essential in understanding the pathophysiology of the disease, and suitable recruitment for clinical trials. Inflammation is a known mediator for neurodegeneration, hence various groups found remarkably high levels of CSF and serum proinflammatory cytokines in PD patients.[21,26] Since their levels are positively associated to the disease severity, they could also be potential biomarker of PD.[27]

Neurodegenerative MDs are broadly classified as synucleinopathy and taupathy. Misfolded Tau proteins (phosphorylated) often aggregates like synuclein proteins and likely to precipitate neurodegeneration. Tau protein levels in CSF can help to discriminate PD and Parkinson's-dementia from AD.[28]

Apart from PD other synucleinopathies are Multiple system atrophy, Lewy Body Dementia, Parkinson's Disease Dementia and Pure Autonomic Failure. Whereas Taupathies are classically Progressive supranuclear palsy, Fronto-Temporal Dementia, Corticobasal Degeneration and Alzheimer's disease.[29] Although, in Alzheimer's disease apart from Tau protein, aggregates of Aβ42 is also found.[28]

Advanced neuroimaging

The second category of biomarker, which is already in clinical practice, is nuclear imaging/SPECT scan. SPECT is the abbreviation for "Single Photon-Emission Computerized Tomography". The tracers for SPECT, specifically Technetium-99m labeled tropane derivative(^{99m}Tc-TRODAT-1), binds to the dopamine transporter (DAT) in the presynaptic membrane and could be used to image the dopaminergic system. It is reported that there is a close relationship between DAT concentrations and striatal dopamine levels.[30] Hence, TRODAT SPECT could provide a useful tool for evaluation of PD. Before the introduction of TRODAT scan, the diagnosis of PD used to depend on clinical evaluation by the practitioner. Even in specialized centers for MDs, a significant percentage of PD patients were reclassified as the disease progressed.[31] A prospective, 3-year longitudinal, multicenter study assessed the agreement of PD diagnosis between clinical scoring methods and DaT-SPECT findings in patients with an uncertain clinical diagnosis. They found that clinical scoring methods could lead to an overdiagnosis of PD, and it was suggested that DaT-SPECT should be implemented to increase the accuracy of diagnosis.[32] However, evidence is still very limited to encourage this expensive test in routine clinical practice, and it should only be restricted for cases that are a diagnostic dilemma. In early stages of PD the striatal neuronal deficiency predominantly affects one side whereas the involvement is usually bilateral in advanced stage.[33] Unlike PD, drug induced Parkinsonism, NPH and vascular Parkinsonism show no abnormality in the uptake of presynaptic tracers in the striatum as none of these conditions affect the presynaptic dopaminergic system. However, atypical Parkinsonisms (MSA, PSP and CBGD) are difficult to distinguish from PD through TRODAT scan.[34] Another imaging modality being used is Positron Emission Tomography (PET) scan. Several different types of PET ligands exist that can assess the integrity of dopaminergic neurons, including ligands that bind to the DAT, vesicular monoamine transporter. Another ligand [18F]-fluorodopa, helps in assessing the activity of striatal aromatic amino acid decarboxylase. All of these ligands show decreased uptake in the caudate and putamen in patients with early PD compared to controls.[35–37] Notably, unlike amyloid-β and Tau for Alzeimer's disease, no validated PET-tracers currently exist to detect alpha synuclein in the brain, despite various development efforts.[38]

Two structural MRI techniques have recently gained popularity in imaging research for PD- Nigrosome and Neuromelanin (NM) imaging. Nigrosomes are small clusters of dopaminergic cells within the dorsal aspect substantia nigra that do not contain calbindin D28K on immunohistochemistry. Five nigrosomes measuring up to a few millimeters in size have been described among which the largest one is located in the dorsal aspect of the tail of substantia nigra pars compacta (SNpc). It is known as nigrosome-1.[39] T2* and susceptibility weighted MRI sequences were used to study Nigrosome or dorsal nigral hyperintensity. Abnormalities in Nigrosome have been consistently observed in PD patients. Reports show it has a reasonably high sensitivity and specificity in differentiating PD from healthy controls, though it is unable to reliably distinguish PD from atypical Parkinsonism (much like TroDAT SPECT scan). In healthy individuals, the dorsal part of SN is hyperintense and forms an ovoid "swallow tail" like appearance; histopathologically this corresponds to the nigrosome containing area. In PD, this area is hypointense on imaging and lacks the characteristic swallow tail appearance. Autopsy specimens of PD brains have showed increase iron accumulation in the corresponding area. Hence, excess iron in the nigrosome is the reason why the dorsal region has reduced signal intensity. Nigrosome imaging appears to be sensitive to early changes in PD, as patients with mild disease and prodromal subjects also demonstrated loss of dorsal nigral hyperintensity.[40]

NM is a dark pigment found in the brain, which is structurally related to melanin. It is found in large quantities in catecholaminergic cells of the SNpc and locus coeruleus, giving a dark color to the structures.[41] T1-weighted spin echo sequence is usually performed for NM imaging. Reduced NM-related signals and/or volume loss in NM containing structures (locus coeruleus, substantia nigra and ventral tegmental area) were found in PD patients. Some studies suggest, PD patients and healthy controls can be quite accurately differentiated through NM sensitive MRI imaging. A longitudinal change of NM signal might be detected in PD, which raises the possibility of using it as a marker to map disease progression and severity, although this area needs more work.

Overall, nigrosome imaging has a better accuracy compared to NM imaging in distinguishing PD from healthy as the latter has a higher degree of variability. Studies with larger sample size are needed to validate its utility in the clinical setting. Besides, distinguishing PD from healthy may not be useful for clinicians, rather differentiating PD from PD mimickers would be more beneficial. So, adequately powered studies with both PD and atypical Parkinsonism are warranted. Additionally, objective quantification of the neurodegeneration through these imaging techniques could be another useful future direction of research.[40] Fig. 11.1 depicts representative images of nigrosome and NM imaging.

Electrophysiology and behavioral studies

Application of electrophysiology is widely used in the diagnosis of seizure disorders and range of neuropathies. However, its utility has not been validated in MD clinics. Estimation of frequency, intensity, and variability of tremor is often used clinically in distinguishing various tremor disorders, especially for differentiating functional from other tremor conditions.[42] Tremor distractibility occurs when the patient is unable to maintain tremor frequency and character over a prolonged period. Entrainment in tremor refers to high coherence of "involuntary tremor" with voluntary activities. Both entrainment and distractibility are properties of psychogenic/functional tremor that can be clinically assessed but need a confirmatory EMG in certain cases.[43] Tremor frequencies are overlapping among various disorders; hence researchers have found diagnostic indices (e.g. tremor stability index) through advanced signal analysis technique.[44] Before these indices are used clinically, it needs validation through properly designed clinical studies.

The utility of EEG recording is questionable in PD patients; EEG recordings in nondemented PD patients are expected to be essentially normal. However, an excessive interregional coherence especially at beta band is noted in them. Dopaminergic treatment or DBS could reverse this synchronous oscillation (this has been described in detail, later).[45,46]

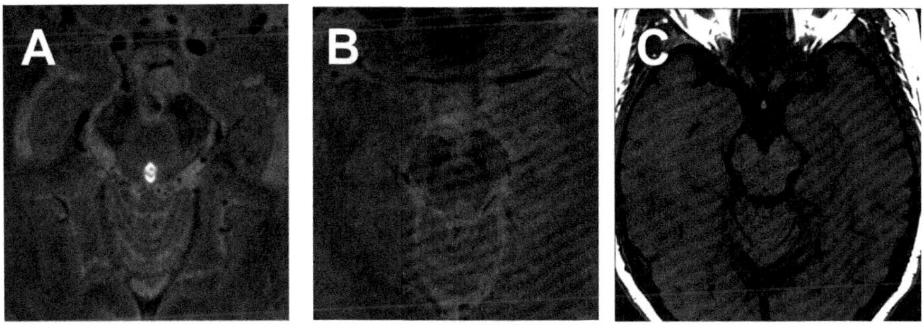

Figure 11.1 Neuromelanin imaging in a control subject, nigrosome imaging in patients with Parkinson's disease and control subjects (A) Showing normal nigrosomes. (B) Showing loss of dorsolateral nigrosomal hyperintensity. (C) MR Spin echo T1 w image showing neuromelanin in bilateral substantia nigra.

Transcranial magnetic Stimulation (TMS) is a widely researched tool in the field of MD. It is a painless way to stimulate cortical neurons. It generates Eddy current from the magnetic field produced by the TMS coils. The outcome measure of the magnetic stimulation over primary motor cortex (M1) could be estimated through electromyogram (EMG). The intensity of muscle contraction as measured through EMG in response to magnetic stimulation is known as motor evoked potential (MEP). The simplest studies performed through TMS generally include estimation of motor threshold (the minimum magnetic stimulation to produce an EMG spike in the corresponding muscle) or quantifying MEP changes with increasing magnetic stimulation intensity (recruitment curve).[47] However, coupling magnetic pulses with afferent nerve stimulation demonstrate changes in the MEP amplitude. For example, MEP is significantly inhibited when median nerve at the wrist is stimulated around 20 ms prior to magnetic pulse over M1. This is an established physiological paradigm called Short Latency Afferent Inhibition (SAI). It is observed that the inhibition is lost or diminished in patients treated with anticholinergic drugs.[48] It probably indicates that SAI is a marker of cortical cholinergic tone. Similarly, various other validated paired pulse paradigms were developed—for example, Short Latency Intracortical Inhibition (SICI), Long Latency Intracortical Inhibition, Intracortical Facilitation. These are predominantly associated with various neurotransmitters.[49]

TMS studies revealed that SAI is reduced in PD dementia, which is partially reversed by cholinesterase inhibitors.[50,51] Surprisingly, reduced cholinergic tone has been found in patients with greater walking difficulty in PD patients, which possibly indicates involvement of nondopaminergic circuits in the motor control of PD.[52] In majority of the studies SICI was found low in PD patients although the reduction does not correlate with the clinical severity.[53] Fig. 11.2 shows the instrument of TMS. It also shows the technique of 2-dimensional (2D) gait assessment which is described in the next section.

Various physiological reflexes are altered in patients with MDs. Brainstem reflexes (e.g., Blink reflex) are diminished more in patients with PSP that in PD.[54] Startle reflex is deranged in patients with freezing of gait.[55] Behavioral studies revealed altered response inhibition in PD patients and patients with dystonia. Recent finding suggests, a measure of motor response inhibition (e.g., ocSSRT) is a sensitive tool to identify even early PD.[56] This paradigm was translated in the Deep Brain Stimulation (DBS) programming clinic and found that it could be used as an adjunctive tool to assess the optimum setting of DBS parameters.[57]

Based on neuroimaging and electrophysiological (TMS) studies, scientists have proposed that the primary motor cortex could have some role in levodopa-induced dyskinesia. Studies like this emphasize the importance of collaborative translational research in understanding the mechanism of not just the disease but also unexplained side effects of commonly used therapies.[58]

Long latency blink reflex (R2 component) is physiologically inhibited by sensory input (e.g. electrical stimulation of median nerve), if administered just before the supraorbital nerve stimulation. Various groups observed that prepulse inhibition of blink reflex is reduced in patients with focal dystonia. Hence it is now proposed

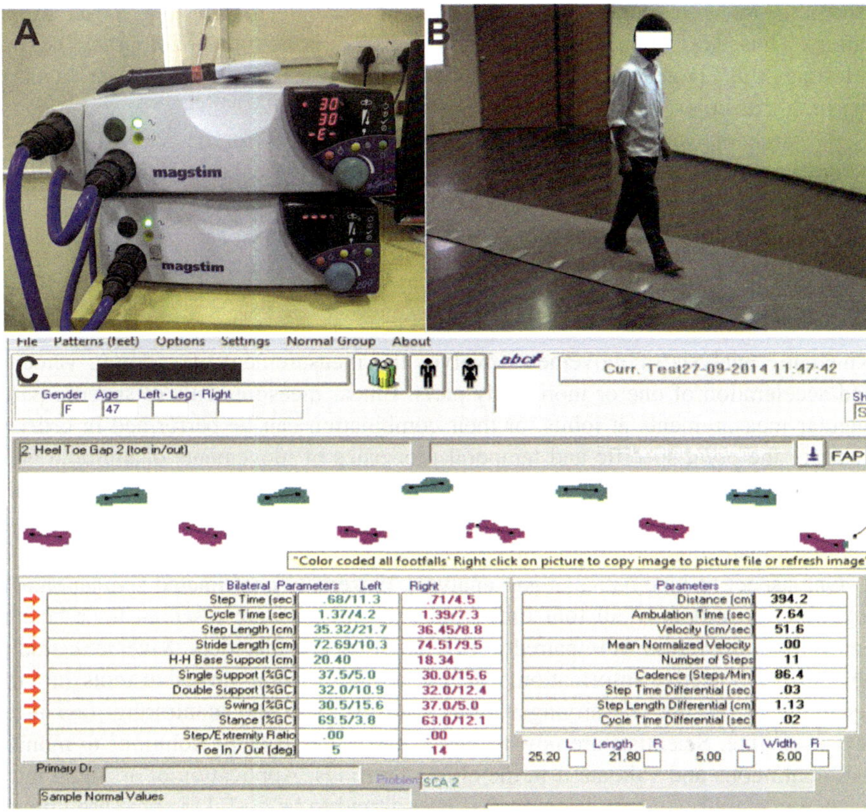

Figure 11.2 Technique of Transcranial Magnetic Stimulation and two-dimensional gait assessment. (A) Transcranial magnetic stimulation instrument. (B) A two-dimensional gait assessment is underway on an electronic walkway (GaitRite). (C) The foot imprints and gait parameters derived from GaitRite.

that the brain stem contributes to the dystonia disease process.[59] Apart from motor responses, sensory processing is affected in patients with dystonia. One of the ways to estimate sensory processing is through measurement of Temporal Discrimination Threshold (TDT). The TDT is the shortest time interval at which a subject can detect that two sensory stimuli are asynchronous. Adult-onset primary torsion dystonia (AOPTD) is the most common form of dystonia which is mostly sporadic but up to 25% of these are familial. Familial AOPTD is inherited in an autosomal dominant fashion with a penetrance as low as 12%–15%. It was observed that unaffected first-degree relatives of such patients with MRI signal changes in the basal ganglia demonstrate a prolonged TDT. A recent study also concluded that the seemingly sporadic AOPTD actually had genetic origin. Abnormal TDTs in patients with sporadic AOPTD and their unaffected first-degree relatives are compatible with an autosomal dominant endophenotype. Most patients with AOPTD have a genetic cause with sporadic cases representing the only manifesting carrier in that family. Thus, TDT appears to be a sensitive marker in both manifesting and nonmanifesting AOPTD gene carriers and therefore of use in genetic studies and counseling in this disorder.[60]

Research on various electrophysiological biomarkers for detection of PD and delineation of various MDs is underway. Many of these have yielded promising results but are yet to be translated in day-to-day clinical practice due to lack of adequately powered studies to establish their clinical relevance.

Kinematics

Kinematic analysis of movement involves the measurement of position, velocity, and acceleration of one or more body parts. Linear measurements of specific parts, angular measurements at joints, or their combination can be performed in order to quantify the point specific and temporal properties of movements of multiple body parts simultaneously. Kinematic methods of analysis have been used to quantify the pattern, severity, or impact of several MDs. These methods have also been applied to discriminate among multiple MDs.[61]

One aspect of kinematics is gait analysis. Extremely coordinated semiautomated rhythmic movement of all four limbs and axial muscles are essential for a healthy gait. This coordination is impaired in PD and other gait disorders. Kinematic studies help to get objective information which could act as non–invasive tools in early screening of disease, classifying various gait disorders and monitoring their progress over time. Scientists are now trying to make user friendly solutions to monitor gait continuous and validate it in the clinical settings. Application of artificial intelligence in classifying gait disorders has been found to be useful in some preliminary studies.[62] The gait assessment could either be three-dimensional (3D) or two-dimensional (2D). In 3D assessments kinematic data from all four limbs and axial structures are available. However, 2D gait analysis is a much more portable tool where, a gait mat embedded with a pressure sensor grid is used for estimation of step length, step time, velocity and numerous other walking parameters for lower limbs only (Fig. 11.2). The combination of kinematics with electrophysiological tools like EMG has been extensively used to study animal models of various MD to identify core features of a disease, similar principles may be used to study these MD in humans.[63]

Progressive impairment of gait and balance is one of the important clinical features in PD.[64] Hence, gait analysis has been widely used in the research involving patients with MDs. PD patients classically walk with short shuffling steps and usually show a stooped posture while standing and walking. Along with this, reduced arm swing is one of the early features of PD.[65] A study on kinematic analysis in PD fallers has shown that there are significant differences in stride characteristics and segmental control between them and PD nonfallers.[66] Perhaps it is important to ensure that these gait characteristics are assessed clinically to facilitate the early identification of PD patients who have a higher propensity of falling. This is helpful for the development of effective intervention strategies to assist fall prevention and personalized care.

A recent publication demonstrated the utility of kinematic parameters as measured through electronic sensors (goniometer etc.) in selection of affected muscles in focal dystonia.[67] Kinematics has also been used to delineate between tics, dystonia, tremor and other MDs in childhood.[61]

Gait and kinematic studies are an important focus in MD research. However, there are number of challenges yet to overcome. Apart from being accurate and precise, the technology should be cost-effective, safe and user friendly. The results should be easily interpreted by the patients' family members and treating physician. This could be an area of future research.

Genetics

In general, PD is often broadly classified as genetic and sporadic. Genetic PD generally follow a monogenic inherence which is either autosomal dominant or autosomal recessive. When it comes to PD, the term monogenic might be a useful oversimplification. The presentation of monogenic PD is quite diverse and it might depend on various nongenetic factors. Studies in the United kingdom attribute genetic PD to be anywhere from 1.4%–17% of the total cases (depending on the number of genes being analyzed and demographics of the population).[68,69] Shifting from candidate protein-screening, researchers are now approaching "-omics" to identify sensitive and specific biomarkers for PD. The principle of "-omics"(genomics for DNA, transcriptomics for RNA, epigenomics for histone or DNA modifications, etc.) research is to find association between suspected DNA, RNA or protein changes in the pathogenesis of a disease with a hypothesis-free approach. PD is quite a heterogenous disease, perhaps deep phenotyping of large dataset with maximum clinical and wet lab data could finally predict the risk of idiopathic/sporadic PD, or infact any MD.[70]

Just like monogenic PD, to describe Parkinson's disease as sporadic (or idiopathic) is also an oversimplification. The genetic basis for some of the so-called idiopathic PD is now known. Through a large-scale genome wide association study (GWAS) the first genetic risk loci for idiopathic PD was identified in 2009. In that study the genetic makeup of 5000 patients were compared with 9000 disease free control subjects. GWAS studies run in a hypothesis free manner, where the entire genome is sequenced instead of assessing candidate genes. Comparing the results with a control group identifies genetic risk factors of PD. In 2019, 90 independent risk factors were identified in another GWAS study with 37,000 patients, 17,000 so-called proxy cases (individuals with a parent with PD) and 1·4 million controls. The GWAS loci identified to date can only explain a fraction of the heritable component of PD.[71] Moreover, molecular studies characterizing these loci have only succeeded in confirming some of the genes as causal for PD. Majority of them are variants of undetermined significance, further highlighting an interplay between genetics, ageing and environmental factors in the pathogenesis of PD.[72]

Despite the completion of more than one study with a large sample size, more GWAS studies in various races and ethnicities are required, as genetic diversity is prevalent in the field of PD. Till now, majority of GWAS studies have been

Table 11.1 Genetic causes of monogenic Parkinson's disease.

Gene protein locus	Mode of inheritance	Clinical features
SNCA (alpha-synuclein)	AD	Young onset PD (YOPD) with atypical features, dementia
Parkin/PARK 2	AR	YOPD with dystonia and dyskinesia
PINK1	AR	YOPD
DJ-1/PARK 7	AR	YOPD
LRKK2 (dardarin)	AD	Resembles IPD, later onset
ATP1342	AR	Early onset, fast progression, atypical signs
PLA2G6	AR	Dystonia with parkinsonism

Many other tentative gene loci are being studied; UCHL1, GIGYF2, HTRA2, PARK16, POLG, FGF20, SNCAIP, NR4A2. *AD*, autosomal dominant; *AR*, autosomal recessive.

conducted in Caucasian population. The lack of diversity in patient population is a major shortcoming in genetic studies, limiting them to have a global impact. GWAS studies are not only useful for risk stratification and developing diagnostic tools but it also gives idea on the mechanism of pathogenesis. With every new gene loci that is identified, some insight regarding the disease-related pathway is improved; even then, it is far from being completely understood.

The Table 11.1 represents established genetic causes of PD.

Among other MDs, the etiology of dystonia can be classified as inherited, acquired, or sporadic. Similarly efforts have been made to identify genetic loci for inherited dystonia syndromes (those with proven genetic origin). DYT-TOR1A was the first dystonia gene mutation to be identified, ever since; many additional genetic forms of dystonia have been described.[73,74] The HD gene (mHTT) was cloned over three decades ago and since then an explosion of research has led to many insights into the normal function of huntingtin protein[75] and the molecular basis of the disease. HD is a polyglutamine disorder with an abnormally high CAG repeat expansion of mHTT gene which leads to an accumulation of toxic protein causing neurodegeneration.[76]

In prevention and treatment

Disease-modifying therapy

Monoclonal antibodies or biologics have been emerged as the primary treatment in various diseases like various cancers, immunological disorders etc. As already discussed, PD is essentially a proteinopathy. Hence, neutralizing antibodies against the offending protein could be an attractive target for disease modifying treatment. It has a potential role in treating patients in its preclinical or prodromal stage as well, thus preventing the onset of the disease. Additionally, prion-like cell to cell transfer of alpha synuclein is thought to be one of the pathogenic mechanisms of dopaminergic cell loss in PD.[77] The famous Braak hypothesis suggests that the synuclein pathology starts in the brainstem and olfactory bulb and it gradually ascends to involve substantia niagra or amygdala in the late stages of PD.[78] As an extension of this hypothesis, some believe that the disease is initiated in the GI tract and gradually ascends to the brain via the vagus nerve (gut-brain axis).[79] In an experiment, alpha synucein positive lewy pathology, was found on autopsy in transplanted nigral fetal cells in patients with PD.[80] This further supports the Braak hypothesis, which rests on the firm idea that synuclein can spread like a prion from one cell to another. This hypothesis has been criticized but enough evidence has almost undoubtably established the spreading nature of synuclein protein. Thus, one of the obvious target for antibody therapy is extracellular alpha synuclein. Few cellular targets have been identified which might be crucial for endocytosis of this protein. The cell surface receptors possibly implicated in aSyn binding are—lymphocyte activation gene-3 (LAG-3), neurexin 1b, A Beta precursor-like protein 1 (APLP1), and connexin-32 (CX32) gap junction proteins have been proposed as possible interactors. Antibodies against these targets might also reduce the risk of prion like spread of offending protein. Hence, antibodies were developed for therapeutic purpose against some of the mentioned targets and also against the N terminal or C terminals of alpha synuclein protein. Some of these drugs are in early phases of clinical development. Pharmaceutical companies like Roche, AstraZeneca, Biogen, AbbVie etc. are currently involved in the development stage of these products. Besides conventional monoclonal antibodies, nanobodies and intrabodies are also developed which can bind to intracellular inclusions unlike conventional peptides. Nanobodies are smaller peptide antibodies, inspired from naturally occurring antibodies in cammels and llamas. An intrabody on the other hand, is an antibody that has been designed to be expressed intracellularly and can be directed to a specific target antigen present in various subcellular locations. The better stability, safety profile and sensitivity of these agents can be attributed to their small size, showing a remarkable potential in therapeutic stratergies.[81] In addition to passive immunization through antibodies, efforts are also made to develop active immunization using recombinant proteins or peptide fragments. Few types of active immunization that have been successful in rodent PD models are—Peptide fragments, Qβ virus-like particles peptide vaccine etc. Active immunization has several advantages as it can be produced in mass scale at a low cost. One of the down sides is that the immune response to active immunization could be variable as most PD patients fall under the geriatric age group and have a suboptimal immune system. Additionally, there is theoretical risk of autoimmune diseases in response to active immunization. For these reasons, current studies mostly focus on passive immunization.[82]

Coming to non PD disorders, disease-modifying treatment for HD is not yet available in the market, but a number of pharmacologic, surgical, and genetic approaches that might delay disease onset or slow progression are being tested.[83] Novel antisense oligonucleotide (ASO) that aims to reduce the synthesis of mutant huntingtin protein by targeting HTT mRNA has shown promising results in the clinical trials and awaiting regulatory approval for clinical use. Experimental therapeutics that target DNA and RNA to modulate protein expression are current targets for HD treatment. Future ASO trials are planned, aiming to improve motor or neuropathological disease manifestations. RNA-based agents require viral delivery, this is advantageous because if developed, it will be a single dose therapy, however long-term toxicity and several other issues must be taken into account. Several of these programs are nearing clinical trials. Zinc-finger transcriptional repressors may reduce huntingtin by targeting DNA without altering it, while CRISPR-Cas9 therapeutics propose to permanently correct the CAG repeats that cause HD in the first place—but these approaches involve viral delivery and bring additional liabilities of their own, so are still in preclinical stages. Monthly intrathecal injections of an ASO directed against mutant huntingtin (*HTT*) were safe and resulted in a dose-dependent decrease in the concentration of mutant huntingtin protein in cerebrospinal fluid compared with placebo. This is one of the promising therapies in HD, 2 open label extension studies on patients who have already taken this drug and a phase 3 double blinded placebo controlled trial are currently underway.[84,85]

Pharmacological therapy

For the last 6 decades, levodopa(l-dopa) is the most effective treatment for PD. Other PD medications introduced after it have not matched its effectiveness. Although the molecule is very effective for majority of the patients with PD, it is quite notorious for causing intolerable adverse effects like dyskinesia, hallucination, somnolence etc. Pulsatile increase in l-dopa concentration in plasma is thought to be associated with dyskinesia. Numerous preclinical studies in MPTP treated primates, showed that treatment with a long-acting dopamine agonist is associated with reduced dyskinesia compared to levodopa.[86] These observations led to several studies of the human basal ganglia demonstrating that in the normal condition striatal dopamine levels are constantly maintained and the administration of l-dopa in a pulsatile manner induces molecular changes in striatal neurons. Conventional pulsatile l-dopa therapy also demonstrated neurophysiologic changes in pallidal output neurons which possibly led to the development of motor complications.[87] Whether continuous dopaminergic therapy has a protective role in neuro-degeneration, could be a topic of future studies.

Hence, efforts have been made to increase the duration and maintain a steady state concentration of l-dopa through various approaches. This is called continuous dopaminergic therapy. The current available options for continuous dopaminergic therapy include L-dopa carbidopa intestinal gel, DBS and apomorphine pump etc.

Levodopa/carbidopa intestinal gel

Duodopa (trade name for levodopa/carbidopa intestinal gel in EU and UK) was approved by the Swedish Medical Products Agency in 2004, followed by the European Medicines Agency for the rest of Europe. The U.S. Food and Drug Administration (FDA) granted the formulation (brand: duopa) as an orphan drug designation in 2015. In the same year, the National Health Service also announced funding and approval of Duodopa in the treatment of PD. Duodopa (Duopa, ABT-SLV187), is a formulation of levodopa and carbidopa developed by AbbVie. It is administered into the intestines at a constant rate for approximately 16-hours as a therapy for patients with advanced PD. It is administered through permanent percutaneous endoscopic gastrostomy with jejunal tube (PEG-J). The dose should be adjusted to an optimal clinical response for the individual patient, which means maximizing the functional ON-time during the day by minimizing the number and duration of OFF episodes (bradykinesia) and minimizing the large fluctuation in ON-time if the patient experiences disabling dyskinesias. Duodopa should be given initially as monotherapy.[88] If required other medicinal products for PD can be taken concurrently. Studies should be undertaken to assess the effectiveness, safety and pharmacoeconomic aspects of duodopa especially in resource poor settings.

Apomorphine pump and pen

Apomorphine is a potent nonergot dopamine agonist derived from morphine. Its narcotic and abuse potential are eliminated. This molecule contains a moiety, homologous to the dopamine molecule, explaining its benefits in PD. Earlier, scientists attributed apomorphine-induced stereotypical behavior in rodents to the direct stimulation of postsynaptic dopamine receptors. However, paradoxical suppression of chorea and dyskinetic movements in humans led to the hypothesis that apomorphine acts additionally on presynaptic receptors apart from postsynaptic dopamine receptors. Eventually, research on apomorphine influencing other pathways like seretoninergic to reduce tremor was published. It is now well known that apomorphine acts on serotonergic and adrenergic receptors. This drug was first tried for PD in 1884 by Dr Edmond Weill, however despite being the first dopamiergic agent, apomorphine suffered decades of neglect and finally began being used clinically since the last 3 decades.[89] It started gaining popularity for PD once its emetic properties could be controlled with prokinetic agents. Apomorphine is currently used for clinical assessment of dopaminergic response in Parkinsonism subjects and for management of moderately advanced PD either as rescue therapy (due to its rapidity of action) or as a continuous dopaminergic stimulation therapy using a pen or a pump, respectively. Gastric motility is reduced in advanced PD, hence oral drugs are partially absorbed. Being the only injectable PD medication, it by passes this issue. Its adverse effects are either due to its mode of administration or pharmacological features. Apart from this apomorphine therapy is suitable for individuals with significant On and Off phenomena, contraindication to DBS surgery due to other comorbidities, peak dose dyskinesia or early morning akinesia.[90]

Its therapeutic effects have been seen in sydenhams chorea, HD, choreoathetosis, orofacial dyskinesia for decades.[89] More recently, a double-blind, randomized, crossover trial showed sustained improvement in chorea without affecting mood symptoms in HD patients. These results need to be confirmed by larger studies, but they already suggest that continuous apomorphine infusion could be another option in certain cases of HD.[91] Recent research on using apomorphine for tourette syndrome is promising and ongoing.[92]

The availability of the drug across countries is not uniform. Hence, data from various population are necessary to ensure safe and effective use of this drug.

Surgical therapy

Deep brain stimulation

Surgical ablation of ventrolateral and ventroanterior nucleus thalamus has been found effective in the treatment of motor symptoms (predominantly tremor) since 1950s. Able Fessard and colleagues (1963) first reported that high frequency (100–120 Hz) electrical stimulation of ventral intermediate nucleus of thalamus could significantly improve symptoms of PD[93]. Perhaps the then poor safety profile and invasiveness made DBS and ablative surgery an unpopular option. Moreover, since pharmacology therapy with levodopa seemed to be very effective, it almost completely sabotaged surgical options initially. It took few decades to understand that l-dopa although effective was not a panacea for PD treatment and its unacceptable adverse effects on long term use were unmasked. Irvine Cooper was the first who utilized chronic DBS for sustained tremor control and Benabid and colleagues paved the way for DBS's worldwide applications (initially in clinical studies and later in clinical routine) almost revitalizing neuro-surgical therapeutics in PD.[94] The concurrent understanding of the neurophysiological basis of MDs helped advance DBS; one worthy of mention is the coherent model of basal ganglia's functional anatomy and its relation to movement development by Albin, DeLong and colleagues. This model was heavily influenced by primate experiments throughout the 1980s, particularly the MPTP model of Parkinsonism.[95,96]

In this backdrop Medronic, Inc. (Minneapolis, MN, USA) released the first commercially available fully implanted DBS system in the mid-1970s for chronic implantation in patients with PD and other MDs. Benabid and Blond and Siegfried independently reported their results of successfully treating advanced PD through high frequency stimulation delivered by modern DBS in 1991. Further primate study by Tipu Aziz and colleagues found Subthalamic nucleus (STN) as an effective surgical target in PD (1991). The work was translated by Pollak et al. within two years (1993) after the seminal primate publication, indicating efficacy in treating PD motor symptoms. Globus pallidus internus (GPi), as another viable target was identified shortly. DBS rapidly gained popularity by the end of that decade and today over 100,000 patients worldwide have received the procedure, it has largely replaced lesioning procedures (pallidotomy and thalamotomy) for treating patients with advanced PD at major medical centers due to its relative safety, minimal invasiveness and entirely reversible physiologic effect.[97]

DBS for motor response of PD has been hypothesized to work in the following way: two downstream physiologic effects are seen in PD neurocircuit. 1. increased STN excitation of the GPi 2. increased GPi inhibition of the thalamus. These, in turn, cause reduced thalamocortical activity. The thalamus is believed to send excitatory stimuli to the cortex, lack of which is believed to mediate akinesia/bradykinesia and rigidity. Pallidotomy reverses the excess pallidal inhibitory effect on the thalamus, improving Parkinsonism by increasing thalamocortical outflow. High-frequency DBS suppresses neuronal activity along with activating efferent fiber pathways that leave the targeted nucleus. Eventhough this is an accepted explanation, DBS's mechanism of action is largely unknown.[98] A long and rich history of research in animals suggests that the PPN may affect locomotion, the startle response, states of arousal, and even reward. Of relevance to gait, PPN has been considered a key component of the mesencephalic locomotor region—an area where electrical stimulation in decerebrated animals can induce locomotion-like activity, whether this concept is of relevance to the therapeutic impact of PPN DBS is still debated.[99]

The modern device is an implantable pulse generator with an integrated battery, which lasts several years or is rechargeable. The battery life depends on the stimulation paradigm and hence the indication. In dystonia patients the battery life is usually shorter than PD as it requires high frequency stimulations compared to majority of PD patients. All the modern DBS devices are fitted with a radiofrequency antenna for communication with external programming devices. These devices are used for adjusting the parameters (frequency, voltage and pulse width), checking impedance and editing the stored data. Currently, manufactures have launched "app" based programmers to gradually replace purpose-built devices which will subsequently communicate through consumer grade communication protocols—like Bluetooth. The stimulation is delivered through transcranial electrodes with multiple contacts. The evidences accumulated so far has helped us to gain insight regarding the clinical outcome of DBS surgery in various targets. Although there is no head to head comparison of STN vs GPi stimulation in PD patients in any adequately powered study, majority of the neurologists and neurosurgeons prefer STN stimulation in PD over GPi. Possibly due to more experience of STN in PD than any other target. Both targets remarkably improve appendicular symptoms but are less effective in alleviating axial symptoms like gait, freezing, speech and balance. Alternative targets like, pedunculopontine nucleus and substantia nigra pars reticulata have been reported to improve the axial symptoms in few pilot studies, recent metanalysis showed disappointing results of PPN stimulation in improving gait and balance in PD patients. It is not clearly understood but almost universally followed that patients with known genetic PD or features of atypical Parkinsonism should be avoided for this procedure. The effectiveness of DBS depends on appropriate patient selection along with postsurgical and periodic programming.

DBS is indeed an expanding field with very high potential. The areas which needs further attention are—understanding its precise mechanism, developing novel technologies and surgical techniques so that it can be offered to the masses. It has been seen that after stimulation the adjacent neurons around the implanted electrode could either increase or reduce firing. Studies have shown that activities preceeded by a beta oscillation are significantly slower than ones which are not preceeded by β oscillations. Synchronous beta oscillations are common manifestation in PD patients. Since bradykinesia is a tell tale sign of PD, several researchers have postulated that DBS desynchronises these oscillation in the beta band, improving motor symptoms in PD patients. More primate experiments are needed for a better understanding of the precise mechanism of DBS. The technologists are developing longer life batteries and energy efficient microprocessors which will be able to deliver complex stimulation paradigms. Closed loop or adaptive DBS is an emerging technology where continuous recording of brain signals will automatically adjust the parameters in real time without manual programming. Teleprogramming using consumer grade communication protocols like WiFi or Bluetooth seem like attractive options, but before implementing them protection against the risk of cybercrime must be ensured. Additionally, use of noninvasive techniques like focussed ultrasound (FUS) or TMS could be developed further as alternative approach for DBS. Table 11.2 describes the MD for which DBS has been routinely studied or used along with the proposed circuits involved. Till now, DBS for PD is usually performed in advanced disease where almost all the patients are essentially refractory to dopaminergic therapy. There are limited experiences on DBS surgery in relatively early cases. Future studies could answer the question with confidence. Currently we only rely on the neurologists' clinical judgment while choosing candidates suitable for the procedure. Whether advance imaging techniques including tractography would be useful in predicting outcomes of DBS surgery, is still a question of debate and an area of future research.

Apart from PD, DBS has become the standard of care for patients with treatment-refractory motor circuit disorders, namely dystonia and essential tremor.

The development of DBS for dystonia has lagged about a decade behind its use in PD. Underlying mechanisms involving DBS effectiveness in dystonia include modification of maladaptive plasticity, progressive motor learning, and altered inhibition and alterations in pathological oscillatory activity in basal ganglia circuitry. Interestingly, although dystonia can recur within minutes to hours after stimulation has been turned off in the early postoperative period, the benefits from stimulation

Table 11.2 Experimental and established targets of DBS for the treatment of patients with movement disorders.

Disorder/disease	Circuits involved	Postulated dysfunction	Deep brain stimulation target (s) in practice/or under study	Stage of study
Parkinson's disease, essential tremor or dystonia	Motor	• Beta and theta oscillations • GPi overactivity • STN overactivity • Neuronal bursting	STN, GPi, GPe, VL thalamus, PPN and spinal cord	Standard of care
Tourette syndrome	Motor and limbic	• Direct pathway hyperactive • Lack of thalamocortical inhibition	SCC, NAcc, habenula and medial forebrain bundle	Phase III trials

Gpi, Internal globus pallidum; *NAcc*, nucleus accumbens; *PPN*, pedunculopontine nucleus; *SCC*, subcallosal cingulate white matter; *STN*, subthalamic nucleus; *VL*, ventrolateral.

that has been administered for several years can persist for days and weeks after cessation. Unlike in PD, DBS in dystonia might, act as a truly disease-modifying treatment. This may justify its use earlier in the course of the disease before poor response to conservative treatment. The most commonly stimulated area in DBS is GPi, although some recent studies on STN show potential as well.[100–102] Essential tremor was the first MD indication for which DBS was approved by the FDA in 1997, after its efficacy was proved in numerous studies, it has become a routine treatment specially since ET commonly requires bilateral modulation making lesional procedures a less suitable option.[103]

MRI-guided focussed ultrasound

MRI-guided FUS is a newer, minimallyf invasive method that uses high-energy ultrasound beams to generate brain lesions. Like conventional lesioning procedures as well as DBS, the potential targets of MRI-guided FUS in patients with PD are the thalamus, subthalamus, and GPi, and predominant symptoms determine the relevant target. All targets for lesioning procedures are unilateral. In MRI guided FUS —the correct target (to lesion) is identified in the first step, the MRI acting as the surgeons eyes, and the US rays as knife that aim to denature protein causing targeted cell death.[104] MRI-guided FUS subthalamotomy is under investigation and has shown promising results preliminarily in patients with PD, but does not yet have regulatory approval for this target yet.[105]

Botulinum toxin

Botulinum toxin (BoNT) injection is the gold standard for the treatment of majority of dystonic conditions. It is a natural toxin isolated from spore forming gram negative bacteria—clostridium botulinum. It was first described by a young German physician, Justinus Kerner, who reported fatal outbreak after intake of bad sausages (botulus, Latin for sausage). In the late twentieth century, Dr Alan B Scott first injected it clinically in the extra ocular muscles of patients with strabismus. Allergan first came up with commercially available BoNT formulation for clinical use in 1989. Since then, more than five pharmaceutical companies launched their products in US for various indications. Approved neurological indications include, various focal dystonia, spasticity, chronic migraine, hemifacial spasm etc. Besides cosmetic indications, bladder conditions and hypersalivation are also labeled for use. Several serotypes of BoNTs are available (A to H). Among these, A and B serotypes are clinically used at present.[106]

BoNT works in the motor end plate and inhibit the release of acetyl choline, which is the primary neurotransmitter responsible for muscle contraction. Membrane proteins endocytose the toxin inside the presynaptic knob, the toxin finally cleaves the SNARE proteins. SNARE proteins are essential for fusion of the vesicles and subsequent release of neurotransmitters in the motor end plate.[107] There are additional mechanisms of BoNT s also. For example, it affects the action of neuropeptides in large dense core vesicles (e.g., calcitonin gene-related peptide,

Substance P) and hence, might be effective in various pain related conditions including migraine.[108]

The identification and isolation of BoNT from nonclostridial sources and synthetic BoNT is a vast field of research, which might have a broader application in future.[109] The high price of BoNT formulations and specialized training required for its application are two main causes of poor acceptability and accessibility of this drug in resource poor countries. Future research focussed to solve these issues is likely to encourage higher number of therapeutic usages of this drug even in developing nations.

Conclusion

This chapter illustrates how the most common MDs are investigated right from their etiology, pathogenesis and diagnosis to their therapies, with an emphasis on contributions made by translational research. Management of MDs has finally metamorphosed from symptomatic treatment to a stage where disease-modifying therapy is now a realistic goal. In this matter, disorders with a more uniform etiology like HD have several promising disease modifying options. On the other hand, more heterogeneous disorders like PD and dystonia will probably require a greater degree of understanding of underlying disease mechanisms to reach this stage.[110] How the advances of clinical research eventually reach the community is an area of equal importance, which must be analyzed by "outcome research". Outcome research believes that even though finding new disease mechanisms and therapies is lucrative, it shouldn't undermine the importance of this knowledge reaching those, for whom it was intended.[2] The latter poses formidable and unique challenges, but mastering them will be even more rewarding. Knowledge of the disease and patient is equally important, as we cannot have a complete grasp on one without being cognizant of the other.

References

1. Cohrs RJ, Martin T, Ghahramani P, Bidaut L, Higgins PJ, Shahzad A. *Translational Medicine Definition by the European Society for Translational Medicine*. Elsevier; 2015.
2. Woolf SH. The meaning of translational research and why it matters. *Jama*. 2008;299:211–213.
3. Mensch J, Oyarzabal J, Mackie C, Augustijns P. In vivo, in vitro and in silico methods for small molecule transfer across the BBB. *J Pharm Sci*. 2009;98:4429–4468.
4. Uenaka T, Satake W, Cha P-C, et al. In silico drug screening by using genome-wide association study data repurposed dabrafenib, an anti-melanoma drug, for Parkinson's disease. *Hum Mol Genet*. 2018;27:3974–3985.
5. Javed H, Nagoor Meeran MF, Azimullah S, Adem A, Sadek B, Ojha SK. Plant extracts and phytochemicals targeting α-synuclein aggregation in Parkinson's disease models. *Front Pharmacol*. 2019;9:1555.

6. Merchant, K.M.; Chesselet, M.-F.; Hu, S.-C.; Fahn, S. Animal models of Parkinson's disease to aid drug discovery and development. In Animal and Translational Models for CNS Drug Discovery, Elsevier: 2008; . 159-205.
7. Bareš M, Kaňovský P, Klajblová H, Rektor I. Intracortical inhibition and facilitation are impaired in patients with early Parkinson's disease: a paired TMS study. *Eur J Neurol.* 2003;10:385–389.
8. Hughes AJ, Daniel SE, Kilford L, Lees AJ. Accuracy of clinical diagnosis of idiopathic Parkinson's disease: a clinico-pathological study of 100 cases. *J Neurol Neurosurg Psych.* 1992;55:181–184.
9. Beach TG, Adler CH. Importance of low diagnostic accuracy for early Parkinson's disease. *Mov Disord.* 2018;33:1551–1554.
10. Shahnawaz M, Tokuda T, Waragai M, et al. Development of a biochemical diagnosis of Parkinson disease by detection of α-synuclein misfolded aggregates in cerebrospinal fluid. *JAMA Neurol.* 2017;74:163–172.
11. Schwarz ST, Afzal M, Morgan PS, Bajaj N, Gowland PA, Auer DP. The 'swallow tail'appearance of the healthy nigrosome–a new accurate test of Parkinson's disease: a case-control and retrospective cross-sectional MRI study at 3T. *PLoS One.* 2014;9: e93814.
12. Chen K-HS, Chen R. Principles of electrophysiological assessments for movement disorders. *J Mov Disord.* 2020;13:27.
13. Billingsley K, Bandres-Ciga S, Saez-Atienzar S, Singleton A. Genetic risk factors in Parkinson's disease. *Cell Tissue Res.* 2018;373:9–20.
14. Wahid F, Begg RK, Hass CJ, Halgamuge S, Ackland DC. Classification of Parkinson's disease gait using spatial-temporal gait features. *IEEE J Biomed Health Inform.* 2015;19:1794–1802.
15. Amur SG, Sanyal S, Chakravarty AG, et al. Building a roadmap to biomarker qualification: challenges and opportunities. *Biomarkers Med.* 2015;9:1095–1105.
16. Killoran A, Biglan K. Biomarkers for Huntington's disease: a brief overview. *J Rare Dis Res & Treat.* 2016;1.
17. Dash D, Mestre TA. Therapeutic update on Huntington's disease: symptomatic treatments and emerging disease-modifying therapies. *Neurotherapeutics.* 2020;1–15.
18. Jellinger KA. The pathomechanisms underlying Parkinson's disease. *Expert Rev Neurother.* 2014;14:199–215.
19. Tsukita K, Sakamaki-Tsukita H, Tanaka K, Suenaga T, Takahashi R. Value of in vivo α-synuclein deposits in Parkinson's disease: a systematic review and meta-analysis. *Mov Disord.* 2019;34:1452–1463.
20. Chahine LM, Beach TG, Brumm MC, et al. In vivo distribution of α-synuclein in multiple tissues and biofluids in Parkinson disease. *Neurology.* 2020;95:e1267–e1284.
21. Chatterjee K, Roy A, Banerjee R, et al. Inflammasome and α-synuclein in Parkinson's disease: a cross-sectional study. *J Neuroimmunol.* 2020;338:577089.
22. Foulds PG, Mitchell JD, Parker A, et al. Phosphorylated α-synuclein can be detected in blood plasma and is potentially a useful biomarker for Parkinson's disease. *FASEB J.* 2011;25:4127–4137.
23. Goldman JG, Andrews H, Amara A, et al. Cerebrospinal fluid, plasma, and saliva in the BioFIND study: relationships among biomarkers and Parkinson's disease features. *Mov Disord.* 2018;33:282–288.
24. Jankovic J, McDermott M, Carter J, et al. Variable expression of Parkinson's disease: a base-line analysis of the DAT ATOP cohort. *Neurology.* 1990;40:1529. 1529.

25. Malec-Litwinowicz M, Plewka A, Plewka D, et al. The relation between plasma α-synuclein level and clinical symptoms or signs of Parkinson's disease. *Neurolo Neurochir Pol.* 2018;52:243–251.
26. Chen H, O'Reilly EJ, Schwarzschild MA, Ascherio A. Peripheral inflammatory biomarkers and risk of Parkinson's disease. *Am J Epidemiol.* 2008;167:90–95.
27. Hall S, Janelidze S, Surova Y, Widner H, Zetterberg H, Hansson O. Cerebrospinal fluid concentrations of inflammatory markers in Parkinson's disease and atypical parkinsonian disorders. *Sci Rep.* 2018;8:1–9.
28. Blennow K. Cerebrospinal fluid protein biomarkers for Alzheimer's disease. *NeuroRx.* 2004;1:213–225.
29. Dugger BN, Dickson DW. Pathology of neurodegenerative diseases. *Cold Spring Harb Perspect Biol.* 2017;9:a028035.
30. Ikeda K, Ebina J, Kawabe K, Iwasaki Y. Dopamine transporter imaging in parkinson disease: progressive changes and therapeutic modification after anti-parkinsonian medications. *Intern Med.* 2019;2489. 2418.
31. Hughes AJ, Daniel SE, Ben-Shlomo Y, Lees AJ. The accuracy of diagnosis of parkinsonian syndromes in a specialist movement disorder service. *Brain.* 2002;125:861–870.
32. Marshall VL, Reininger CB, Marquardt M, et al. Parkinson's disease is overdiagnosed clinically at baseline in diagnostically uncertain cases: a 3-year European multicenter study with repeat [123I] FP-CIT SPECT. *Mov Disord.* 2009;24:500–508.
33. Huang W-S, Lin S-Z, Lin J-C, Wey S-P, Ting G, Liu R-S. Evaluation of early-stage Parkinson's disease with 99mTc-TRODAT-1 imaging. *J Nucl Med.* 2001;42:1303–1308.
34. Vlaar AM, van Kroonenburgh MJ, Kessels AG, Weber WE. Meta-analysis of the literature on diagnostic accuracy of SPECT in parkinsonian syndromes. *BMC Neurol.* 2007;7:1–13.
35. Suh M, Im JH, Choi H, Kim HJ, Cheon GJ, Jeon B. Unsupervised clustering of dopamine transporter PET imaging discovers heterogeneity of parkinsonism. *Hum Brain Mapp.* 2020;41:4744–4752.
36. Jokinen P, Helenius H, Rauhala E, Brück A, Eskola O, Rinne JO. Simple ratio analysis of 18F-fluorodopa uptake in striatal subregions separates patients with early Parkinson disease from healthy controls. *J Nucl Med.* 2009;50:893–899.
37. Hsiao T, Weng Y-H, Hsieh C-J, et al. Correlation of Parkinson disease severity and 18F-DTBZ positron emission tomography. *JAMA Neurol.* 2014;71:758–766.
38. Merchant KM, Cedarbaum JM, Brundin P, et al. A proposed roadmap for Parkinson's disease proof of concept clinical trials investigating compounds targeting alpha-synuclein. *J Parkinson's Dis.* 2019;9:31–61.
39. Li P-J, Jin T, Luo D-H, et al. Effect of prolonged radiotherapy treatment time on survival outcomes after intensity-modulated radiation therapy in nasopharyngeal carcinoma. *PLoS One.* 2015;10:e0141332.
40. Pavese N, Tai YF. Nigrosome imaging and neuromelanin sensitive MRI in diagnostic evaluation of parkinsonism. *Mov Disord Clin Pract.* 2018;5:131–140.
41. Fedorow H, Tribl F, Halliday G, Gerlach M, Riederer P, Double K. Neuromelanin in human dopamine neurons: comparison with peripheral melanins and relevance to Parkinson's disease. *Prog Neurobiol.* 2005;75:109–124.
42. Vial F, Kassavetis P, Merchant S, Haubenberger D, Hallett M. How to do an electrophysiological study of tremor. *Clin Neurophysiol Pract.* 2019;4:134–142.
43. Pal PK. Electrophysiologic evaluation of psychogenic movement disorders. *J Mov Disord.* 2011;4:21.

44. di Biase L, Brittain J-S, Shah SA, et al. Tremor stability index: a new tool for differential diagnosis in tremor syndromes. *Brain*. 2017;140:1977–1986.
45. George JS, Strunk J, Mak-McCully R, Houser M, Poizner H, Aron AR. Dopaminergic therapy in Parkinson's disease decreases cortical beta band coherence in the resting state and increases cortical beta band power during executive control. *NeuroImage: Clin*. 2013;3:261–270.
46. Hirschmann J, Özkurt TE, Butz M, et al. Differential modulation of STN-cortical and cortico-muscular coherence by movement and levodopa in Parkinson's disease. *Neuroimage*. 2013;68:203–213.
47. Klomjai W, Katz R, Lackmy-Vallée A. Basic principles of transcranial magnetic stimulation (TMS) and repetitive TMS (rTMS). *Ann Phys Rehabil Med*. 2015;58:208–213.
48. Di Lazzaro V, Oliviero A, Profice P, et al. Muscarinic receptor blockade has differential effects on the excitability of intracortical circuits in the human motor cortex. *Exp Brain Res*. 2000;135:455–461.
49. Di Pino G, Pellegrino G, Assenza G, et al. Modulation of brain plasticity in stroke: a novel model for neurorehabilitation. *Nat Rev Neurol*. 2014;10:597.
50. Celebi O, Temuçin ÇM, Elibol B, Saka E. Short latency afferent inhibition in Parkinson's disease patients with dementia. *Mov Disord*. 2012;27:1052–1055.
51. Di Lazzaro V, Oliviero A, Pilato F, et al. Neurophysiological predictors of long term response to AChE inhibitors in AD patients. *J Neurol, Neurosurg Psych*. 2005;76:1064–1069.
52. Rochester L, Yarnall AJ, Baker MR, et al. Cholinergic dysfunction contributes to gait disturbance in early Parkinson's disease. *Brain*. 2012;135:2779–2788.
53. Ni Z, Bahl N, Gunraj CA, Mazzella F, Chen R. Increased motor cortical facilitation and decreased inhibition in Parkinson disease. *Neurology*. 2013;80:1746–1753.
54. Szmidt-Salkowska E, Gawel M, Jamrozik Z, Salkowska-Wanat J, Gawel D, Kaminska A. Diagnostic value of blink reflex in multisystem atrophy, progressive supranuclear palsy and Parkinson disease. *Neurolo Neurochir Pol*. 2016;50:336–341.
55. Nonnekes J, Geurts AC, Nijhuis LBO, et al. Reduced StartReact effect and freezing of gait in Parkinson's disease: two of a kind? *J Neurol*. 2014;261:943–950.
56. Choudhury S, Roy A, Mondal B, et al. Slowed movement stopping in Parkinson's disease and focal dystonia is improved by standard treatment. *Sci Rep*. 2019;9:1–9.
57. Roy A, Choudhury S, Basu P, Baker MR, Baker SN, Kumar H. Stop Signal Reaction Time measured with a portable device validates optimum STN-DBS programming. *Brain Stimul*. 2020;13:1609–1611.
58. Rothwell JC, Obeso JA. Can levodopa-induced dyskinesias go beyond the motor circuit? *Brain*. 2015;138:242–244.
59. Choudhury S, Singh R, Chatterjee P, et al. Abnormal blink reflex and intermuscular coherence in Writer's cramp. *Front Neurol*. 2018;9:517.
60. Bradley D, Whelan R, Walsh R, et al. Temporal discrimination threshold: VBM evidence for an endophenotype in adult onset primary torsion dystonia. *Brain*. 2009;132:2327–2335.
61. Singer HS, Mink J, Gilbert DL, Jankovic J. *Movement Disorders in Childhood*. Academic Press:; 2015.
62. Prakash C, Kumar R, Mittal N. Recent developments in human gait research: parameters, approaches, applications, machine learning techniques, datasets and challenges. *Artif Intell Rev*. 2018;49:1–40.
63. Scholle HC, Jinnah H, Arnold D, et al. Kinematic and electromyographic tools for characterizing movement disorders in mice. *Mov Disord*. 2010;25:265–274.

64. Raccagni C, Nonnekes J, Bloem BR, et al. Gait and postural disorders in parkinsonism: a clinical approach. *J Neurol.* 2019;1–8.
65. Mirelman A, Bernad-Elazari H, Thaler A, et al. Arm swing as a potential new prodromal marker of Parkinson's disease. *Mov Disord.* 2016;31:1527–1534.
66. Cole MH, Silburn PA, Wood JM, Worringham CJ, Kerr GK. Falls in Parkinson's disease: kinematic evidence for impaired head and trunk control. *Mov Disord.* 2010;25:2369–2378.
67. Samotus O, Lee J, Jog M. Personalized botulinum toxin type A therapy for cervical dystonia based on kinematic guidance. *J Neurol.* 2018;265:1269–1278.
68. Tan MM, Malek N, Lawton MA, et al. Genetic analysis of Mendelian mutations in a large UK population-based Parkinson's disease study. *Brain.* 2019;142:2828–2844.
69. Neurol A. Frequency of known mutations in early onset PD; implication for genetic counseling: the CORE-PD study. *Arch Neurol.* 2010;67:1116–1122.
70. Diaz-Ortiz ME, Chen-Plotkin AS. Omics in neurodegenerative disease: hope or hype? *Trends Genet.* 2020;36:152–159.
71. Blauwendraat C, Nalls MA, Singleton AB. The genetic architecture of Parkinson's disease. *Lancet Neurol.* 2020;19:170–178.
72. Pang SY-Y, Ho PW-L, Liu H-F, et al. The interplay of aging, genetics and environmental factors in the pathogenesis of Parkinson's disease. *Transl Neurodegener.* 2019;8:1–11.
73. Albanese A, Bhatia K, Bressman SB, et al. Phenomenology and classification of dystonia: a consensus update. *Mov Disord.* 2013;28:863–873.
74. Charlesworth G, Bhatia KP, Wood NW. The genetics of dystonia: new twists in an old tale. *Brain.* 2013;136:2017–2037.
75. Caviston JP, Holzbaur EL. Huntingtin as an essential integrator of intracellular vesicular trafficking. *Trends Cell Biol.* 2009;19:147–155.
76. Rubinsztein DC, Carmichael J. Huntington's's disease: molecular basis of neurodegeneration. *Expert Rev Mol Med.* 2003;5:1.
77. Masuda-Suzukake M, Nonaka T, Hosokawa M, et al. Prion-like spreading of pathological α-synuclein in brain. *Brain.* 2013;136:1128–1138.
78. McCann H, Cartwright H, Halliday GM. Neuropathology of α-synuclein propagation and braak hypothesis. *Mov Disord.* 2016;31:152–160.
79. Breen DP, Halliday GM, Lang AE. Gut–brain axis and the spread of α-synuclein pathology: vagal highway or dead end? *Mov Disord.* 2019;34:307–316.
80. Chu Y, Kordower JH. Lewy body pathology in fetal grafts. *Ann N Y Acad Sci.* 2010;1184:55–67.
81. Messer A, Butler DC. Optimizing intracellular antibodies (intrabodies/nanobodies) to treat neurodegenerative disorders. *Neurobiol Dis.* 2020;134:104619.
82. Castonguay A-M, Gravel C, Lévesque M. Treating Parkinson's disease with antibodies: previous studies and future directions. *J Parkinson's Dis.* 2020;1–22.
83. Wild EJ, Tabrizi SJ. Therapies targeting DNA and RNA in Huntington's's disease. *Lancet Neurol.* 2017;16:837–847.
84. Shannon KM. Recent advances in the treatment of Huntington's's disease: targeting DNA and RNA. *CNS Drugs.* 2020;34:219–228.
85. Tabrizi SJ, Leavitt BR, Landwehrmeyer GB, et al. Targeting huntingtin expression in patients with Huntington's's disease. *N Engl J Med.* 2019;380:2307–2316.
86. Tabrizi SJ, Flower MD, Ross CA, Wild EJ. Huntington's disease: new insights into molecular pathogenesis and therapeutic opportunities. *Nat Rev Neurol.* 2020;16:529–546.

87. Barroso Chinea P, Bezard E. Basal ganglia circuits underlying the pathophysiology of levodopa-induced dyskinesia. *Front Neuroanat.* 2010;4:131.
88. AbbVie Inc., N.C., IL. Highlights of prescribing information: duopa (carbidopa and levodopa) enteral suspension. Administration, U.S.F.a.D., Ed.; January 2015.
89. Andén NE, Rubenson A, Fuxe K, Hökfelt T. Evidence for dopamine receptor stimulation by apomorphine. *J Pharm Pharmacol.* 1967;19:627–629.
90. Prashanth L, Jaychandran R, Seetharam R, Iyer RB. Apomorphine: the initial Indian experience in relation to response tests and pumps. *Ann Indian Acad Neurol.* 2020;23:20.
91. Vitale C, Marconi S, Di Maio L, et al. Short-term continuous infusion of apomorphine hydrochloride for treatment of Huntington's's chorea: a double blind, randomized cross-over trial. *Mov Disord.* 2007;22:2359–2364.
92. Feinberg M, Carroll BJ. Effects of dopamine agonists and antagonists in Tourette's disease. *Arch Gen Psych.* 1979;36:979–985.
93. Zangiabadi N, Ladino LD, Sina F, Orozco-Hernández JP, Carter A, Téllez-Zenteno JF. Deep brain stimulation and drug-resistant epilepsy: a review of the literature. Front Neurol. 2019;Jun 6(10):601.
94. Hartmann CJ, Fliegen S, Groiss SJ, Wojtecki L, Schnitzler A. An update on best practice of deep brain stimulation in Parkinson's disease. *Therap Adv Neurol Disord.* 2019;12. 1756286419838096.
95. Burns RS, Chiueh CC, Markey SP, Ebert MH, Jacobowitz DM, Kopin IJ. A primate model of parkinsonism: selective destruction of dopaminergic neurons in the pars compacta of the substantia nigra by N-methyl-4-phenyl-1, 2, 3, 6-tetrahydropyridine. *Proc Natl Acad Sci.* 1983;80:4546–4550.
96. Albin RL, Young AB, Penney JB. The functional anatomy of basal ganglia disorders. *Trends Neurosci.* 1989;12:366–375.
97. Pycroft L, Stein J, Aziz T. Deep brain stimulation: an overview of history, methods, and future developments. *Brain Neurosci Adv.* 2018;2. 2398212818816017.
98. Chou KL, Tarsy D. Device-assisted and lesioning procedures for Parkinson disease.
99. Thevathasan W, Debu B, Aziz T, et al. Pedunculopontine nucleus deep brain stimulation in Parkinson's disease: a clinical review. *Mov Disord.* 2018;33:10–20.
100. Lozano AM, Lipsman N, Bergman H, et al. Deep brain stimulation: current challenges and future directions. *Nat Rev Neurol.* 2019;15:148–160.
101. Kleiner-Fisman G, Liang GSL, Moberg PJ, et al. Subthalamic nucleus deep brain stimulation for severe idiopathic dystonia: impact on severity, neuropsychological status, and quality of life. *J Neurosurg.* 2007;107:29–36.
102. Kupsch A, Benecke R, Müller J, et al. Pallidal deep-brain stimulation in primary generalized or segmental dystonia. *N Engl J Med.* 2006;355:1978–1990.
103. Singh M, Agrawal M. Deep brain stimulation for tremor and dystonia. *Neurol India.* 2020;68:187.
104. Schlesinger I, Sinai A, Zaaroor M. MRI-guided focused ultrasound in Parkinson's disease: a review. *Parkinson's Dis.* 2017;2017.
105. Martínez-Fernández R, Máñez-Miró JU, Rodríguez-Rojas R, et al. Randomized trial of focused ultrasound subthalamotomy for Parkinson's disease. *N Engl J Med.* 2020;383:2501–2513.
106. Jabbari B. *Botulinum Toxin Treatment.* Springer; 2018.
107. Jabbari B. *Basics of structure and mechanisms of function of botulinum toxin-how does it work?* Botulinum Toxin Treatment. Springer; 2018:11–17.

108. Cutrer F, Sandroni P, Wendelschafer-Crabb G. Botulinum toxin treatment of cephalalgia alopecia increases substance P and calcitonin gene-related peptide-containing cutaneous nerves in scalp. *Cephalalgia*. 2010;30:1000–1006.
109. Webb RP. Engineering of botulinum neurotoxins for biomedical applications. *Toxins*. 2018;10:231.
110. Klein C, Krainc D, Schlossmacher MG, Lang AE. Translational research in neurology and neuroscience 2011: movement disorders. *Arch Neurol*. 2011;68:709–716.

Epilepsy and translational research

12

Mariangela Panebianco
Department of Molecular and Clinical Pharmacology, Institute of Translational Medicine, University of Liverpool, Clinical Sciences Centre for Research and Education, Liverpool, United Kingdom

Introduction

Epilepsy is a disorder in which recurrent and spontaneous seizures are caused by synchronous high-frequency discharges of neuronal populations in the brain. Epilepsy affects over 50 million people worldwide and is thus one of the most common chronic neurological conditions. Epilepsy has numerous causes, each reflecting underlying brain dysfunction.[1] A seizure can be conceptualized as occurring when there is distortion of the normal balance between excitation and inhibition in the brain.[2] A seizure provoked by a reversible insult (e.g., fever, hypoglycemia) does not fall under the definition of epilepsy, because it is a short-lived secondary condition, not a chronic state. "Epilepsy syndrome" refers to a group of clinical characteristics that consistently occur together, with similar seizure type, age of onset, electroencephalogram (EEG) findings, triggering factors, genetics, natural history, prognosis, and response to antiepileptic drugs (AEDs). The most recent International League Against Epilepsy classification of epileptic seizures and epilepsies was published in 2010.[3] Epilepsy is one of the most disabling neurologic conditions, yet we have an incomplete understanding of the detailed pathophysiology and thus treatment rational for much of the epilepsy. Underneath the exuberant behavioral and electroencephalographic expression of epileptic seizures, there are several molecular, metabolic, cellular, and functional changes in brain activity that strongly affect the patient's brain function. Epilepsy is characterized not only by recurrent seizures but also by a two- to threefold increase in mortality, stigma, psychobehavioral comorbidity, and decreased social participation and quality of life. At the moment we are not able to prevent epileptogenesis (the development of epilepsy following an initial precipitating injury such as a febrile seizure of traumatic brain injury), and there are very few disease-modifying treatments. Despite the availability of more than 15 new AEDs, the proportion of patients with drug-resistant epilepsy has remained constant at about 20%–30%.[4] Some studies suggest that the recurrence of seizures may be responsible for the progressive neuronal damage, and some histopathological evidence in patients with epilepsy correlates the occurrence of drug-refractory seizures with lower hippocampal neuronal density.[5] There is evidence that some types of epilepsy progress over time, and an

Perioperative Neuroscience. DOI: https://doi.org/10.1016/B978-0-323-91003-3.00015-5

important part of this knowledge has derived from neuroimaging studies. Different authors have demonstrated structural damage more pronounced in individuals with a longer duration of epilepsy, and others have been able to quantify this progression over time.[6]

Diagnostic evaluation

The techniques applied to evaluate patients with epilepsy and their seizures, such as medical history, seizure semiology, genetic studies, scalp and invasive video-EEG, brain stimulation, neuroimaging, and neuropsychological, social and psychiatric evaluations are fundamental strategies for finding the origins of the seizures and to assess the cognitive and social consequences of epilepsy.[7] EEG and magnetic resonance imaging (MRI) are currently the most widely used techniques, while other methods including molecular biomarkers measured in blood and cerebrospinal fluid are increasingly evaluated.

History and neurologic examination

Important historical features include the clinical context in which the seizure occurred, including premonitory signs, details of seizure itself, such as phenomenology, responsiveness, focal features, and the postictal state. Further, inquire centers guide the nature and extent of the evaluation and determine treatment and prognosis. The neurological examination assesses focal signs that might implicate or localize cerebral pathology.[8]

Neurophysiological examinations

An EEG is a recording of the brain's electrical activity. Electroencephalography biomarkers in daily routine involve the localization of interictal epileptiform discharges and seizure detection.[9] Quantitative EEG (qEEG) is a powerful method to evaluate the frequency composition of EEG activity. It can identify brain areas where there is too much or too little EEG activity compared to the normal database.[10]

Neuroimaging techniques

MRI is the most important neuroimaging tool for the evaluation of epilepsy. It can image substrate responsible for the epileptic process and reveal signs of hippocampal sclerosis, malformations of cortical development, tumors, or other underlying conditions. MRI permits in vivo quantification of brain volume, and has many applications in cognitive, clinical, and comparative neuroscience. We consider the estimation of the volume of a brain structure that is not so well isolated, namely the hippocampus. This structure is of particular interest as it may be routinely estimated in a medical setting in the context of presurgical evaluation in patients with clinical evidence of temporal lobe epilepsy.[11] Malformations of cortical development, including focal cortical dysplasia, periventricular nodular heterotopia, and polymicrogyria, are common causes of refractory epilepsy, and their detection is crucial for successful epilepsy surgery. In contrast to conventional MRI techniques which are based on the acquisition of data with different contrast weighting, the goal of quantitative MRI (qMRI) is the mapping of specific tissue parameters, such as the relaxation times, T1 and T2, and the proton density.[12]

Other neuroimaging techniques

MRI T2 relaxometry is a technique that allows the quantification of signal intensity in MRI scans and can help to improve the detection of structural brain pathologies such as hippocampal sclerosis.[13] MRI spectroscopy allows the observation of cerebral metabolites, and many studies in epilepsy have shown that it can help to demonstrate neuronal damage that correlates with clinical electroencephalographic lateralization in temporal left epilepsy. Some cross-sectional studies using MRI spectroscopy have demonstrated correlation of seizure frequency and epilepsy duration with a reduction in the N-acetylaspartate-to-creatine ratio in patients with medically intractable temporal epilepsy.[14] Positron emission tomography (PET) provides images of different biochemical functions in the human brain; and abnormal glucose metabolism has been associated with the epileptic focus. PET demonstrated dynamic changes of cortical glucose hypometabolism in children with intractable nonlesional partial epilepsy, and these changes were treated to the frequency of seizures.[15] In the last two decades, with the emergence of more sophisticated scans and quantitative analysis, the literature about progressive epilepsy damage and MRI techniques has grown. For other neuroimaging techniques, such as PET or single photon emission computed tomography, the literature regarding the progressive damage of epilepsies is not so vast.[16]

Metabolic evaluation

In metabolic disorders, seizures are typically accompanied by other abnormalities, such as developmental delay, unexplained vomiting, or coma. In neonatal seizures, a metabolic evaluation is mandatory, including a screen of serum amino acids and urine organic acids, and blood lactate to screen for mitochondrial disease. In addition to its more common use, the evaluation of central nervous system infection, cerebrospinal fluid can be analyzed for glucose transporter defects and rare neurotransmitter defects.[17,18]

Genetic testing

Genetic testing is available for several single genes as well as complex genetic disorders.[19] A basic karyotype can be performed to evaluate for a chromosomal anomaly, especially in a patient with dysmorphic features. If a specific syndrome is suspected, an epilepsy panel of selected genes can be ordered. Comparative genomic hybridization microarray evaluates targeted chromosomal regions for copy number variants. When a genetic diagnosis is highly suspected, but other work up is unrevealing, the clinician can consider whole exome sequencing of the patient and parents, a technique with rapidly expanding clinical use, especially in epileptic encephalopathies of unknown etiology[20] (Table 12.1).

Treatment

Drugs used to treat epilepsy work by decreasing the electrical activity of the brain, either by preventing neuronal depolarization by blocking sodium channels or calcium channels, enhancing potassium channel function, inhibiting excitation mediated by the neurotransmitter glutamate, or promoting inhibition mediated by gamma aminobutyric acid.[21] A neurologist decision to begin one seizure medication over another is based on seizure type, age, other medical conditions, and potential side-effect profile. As a general principle, medication should be started at a low dose to avoid side effects. Dosage increase can be performed at regular intervals, if needed. The goal is to control seizures with the lowest dose. When a first drug fails, most clinicians will choose to add on a second drug, later deciding whether or not to withdraw the initial medication. As a result of their mechanism of action, all seizure medications have central nervous system side effects. If medications fail to control seizures, other options include dietary therapy (ketogenic diet), vagus nerve stimulation (VNS), immune therapy, resective epilepsy surgery (lesionectomy, hemispherotomy), and palliative epilepsy surgery (intracranial stimulation therapy, callosotomy). The ketogenic diet has been investigated as treatment for epilepsy

Table 12.1 Old and new challenges of diagnostic in epilepsy.

Challenges
Old challenges
History and neurologic examination
EEG and quantitative EEG
MRI and quantitative MRI
MRI T2 relaxometry
MRI spectrometry
PET
Metabolic evaluation
Genetic testing
New challenges
DTI/tractography
Functional MRI
Neuroethological method
Innate immunity/cerebral microdialysis

DTI, Diffusion tensor imaging; *EEG*, electroencephalography; *MRI*, magnetic resonance imaging; *PET*, positron emission tomography.

since 1920s, but only more recently have the first randomized clinical trials been reported.[22] The main strategy for focal modification of brain tissue to treat epilepsy is surgical ablation. Although this can be effective, the destruction of tissue severely limits this approach, as it is important to avoid eloquent cortex. The role of immune therapy for refractory epilepsy is still being defined. VNS is a neuromodulatory treatment that is used as an adjunctive therapy for treating people with medically refractory epilepsy. VNS consists of chronic intermittent electrical stimulation of the vagus nerve, delivered by a programmable pulse generator.[4] Lifestyle adjustment is a crucial aspect of epilepsy management. Optimizing sleep, improving medication compliance, and reducing stress can significantly improve epilepsy outcome. Neuroprotective and antiepileptogenic approaches have been extensively explored for preventing and treating epilepsy. Although the most commonly used therapeutic approach to control seizures is pharmacological, nonpharmacological therapies including complementary and alternative medicine are often used by people with epilepsy. Acupuncture, botanical/herbals, chiropractical care, magnet therapy, prayers, stress management, and yoga are frequently used.[23,24] Complementary therapies for preventing or treating epilepsy have been extensively used. Considering that exercise can exert beneficial actions such as reduction of seizure susceptibility, reduction of anxiety and depression, and consequently, improvement of quality of life of individuals with epilepsy; exercise can be a potential candidate as nonpharmacological treatment of epilepsy.[25] Currently available AEDs reduce the likelihood of seizure occurrence but do not modify the underlying disease process. The genetic causes of a number of monogenic epilepsies have been discovered recently. Understanding the pathophysiological consequences of the identified mutations increasingly allows the initiation of therapies that modify the effect caused by the specific mutation (precision medicine). This approach often includes repurposing of already available drugs that are approved for the treatment of different diseases.[26]

Future perspective of translational research

Considerable progress has been made in the diagnosis and treatment of epilepsy, but there is much work remaining, that is, the main goals of translational research. The natural history of many different types of epilepsies may be related to the etiology and initial epileptogenic process but may also be associated with many distinct individual characteristics concerning genetic background and exposure to environmental factors. Only with an appropriate multivariate analysis we will be able to have a better view of what is the weight of each intrinsic and extrinsic characteristic for individuals with epilepsy. With the advances in neuroimaging techniques, in vivo and noninvasive structural and functional studies will help to improve the understanding of epilepsy progression. In the last decade, an extensive number of refined quantitative MRI postprocessing tools have emerged. Diffusion tensor imaging/tractography can study the white matter integrity through the evaluation of water diffusion. Diffusion tensor imaging abnormalities have been described in malformations of cortical development and idiopathic generalized epilepsies, but most of studies have focused so far on temporal left epilepsy.[27]

Functional MRI (fMRI) allows the measurement of brain activity by detecting signal changes associated with changes in blood flow and blood oxygenation. It has been used as an important clinical tool for evaluation of memory performance and to map eloquent areas for surgical planning. This capacity of a noninvasive evaluation of hemodynamic changes associated with neural activity allows the possibility for evaluating different aspects related to the epilepsy progression. Multimodal neuroimaging techniques, such as the use of concomitant EEG and fMRI (EEG–fMRI), may be able, in the future, to add more knowledge through a refined noninvasive evaluation. Electroencephalography–fMRI has been proven to help in the identification of the epileptogenic focus in focal refractory epilepsies as well as to evaluate networks involved in different types of epilepsies.[28] A relatively new proposal is to apply neuroethological methods to evaluate epileptic seizures. Neuroethology is a combination of ethology (the comparative study of behavior) and neurophysiology or neurobiology (the study of central nervous system functioning). Neuroethological analysis has been successfully applied and validated in animal models of epilepsy and compulsive behavior and in some types of human epilepsy. The approach is based on a detailed examination of specific behaviors or includes all behavioral items displayed by the subject during the period of observation. Each sign or behavior is recorded and a statistical analysis provides the frequency, duration, and interaction between pairs of behaviors that can be displayed as flowcharts. The neuroethological analysis ends with the description of the flowcharts, displaying all behaviors presented during the period. Some behaviors have localizing and lateralizing value, and their interactions are useful in explaining the brain regions associated with the seizures.[29] Novel approaches and future perspectives for personalized translational research included new developments and ideas in a range of experimental and clinical areas such as microRNA (miRNA) as treatment targets, functional characterization of genetic variants in human cell models and rodent organotypic slice cultures, personalized treatment approaches for monogenic epilepsies, computational modeling for target and biomarker identification.[30] Mechanisms of innate immunity play an important role in the development of acquired epilepsies. Components of the innate immune system, for example, cytokines or Toll-like receptors, can mediate tissue remodeling, leading to network changes that eventually result in increased excitation and synchronization as well as reduced inhibition. In addition, immune mediators also have a direct influence on neuronal excitability via posttranslational modification of ion channels.[31] Such findings highlight the need for personalized approaches in both basic research and patient treatment above and beyond the current standard, general methods. One tool for personalization is cerebral microdialysis that allows repetitive sampling of extracellular molecules at consecutive time points within the same individual. Applying a stereotaxic approach, the target area can be targeted with extreme precision. This technique is used to establish the time course of hippocampal cytokine release in a rat model of mTLE-HS over the course of several months. These results will be helpful for precise timing of immunomodulatory interventions for epilepsy therapy[32] (Table 12.1). An important mechanism to control excitatory synapse function is homeostatic scaling, which regulates synaptic strength in a manner opposite to the stimulus to counteract potentially detrimental changes in the excitatory/inhibitory balance. Homeostatic downscaling in response to chronic overexcitation of networks has been discussed as a pathophysiological mechanism in the context of epilepsy, but the underlying gene regulatory programs are little understood. One hypothesis is that miRNAs, a large class of small regulatory noncoding RNAs, play an important role in the regulation of homeostatic downscaling during epilepsy. More recently, unbiased screening for miRNAs regulated during synaptic downscaling in vitro using small RNA sequencing have been performed; eight miRNAs identified that are upregulated by chronic network activation.[33] miRNAs might therefore be used as biomarkers for a precise classification of different types of epilepsy and to predict therapy outcome. The knowledge about specific miRNA profiles in epilepsy might further inform potential therapeutic strategies that target miRNAs or related pathways. Large international efforts catalyzed by the ongoing progress in epilepsy genetics will produce extended cohorts of patients with similar mutations. Thus personalized medicine is based on precise phenotyping that allows the identification of well-defined groups of patients that are amenable to specific treatment approaches. The recognition of relevant and treatable factors in these group allows the development of tailored therapeutic approaches. Even with the advent of whole genome screening, the complexity of the interaction between genetic, environmental, and stochastic processes means that deep phenotyping still plays a critical role.[34,35] Personalized therapies using new technology are currently still at the preclinical stage. They are termed designer receptor exclusively activated by designer drugs, a class of chemogenetically engineered proteins that permit spatial and temporal control of G protein signaling in vivo; and optogenetics, a method that uses light to modulate molecular events in a targeted manner in living cells or organisms and relies on the use of genetically encoded proteins that change conformation in the presence of light to alter cell behavior by changing the membrane voltage potential of excitable cells. In contrast to resective surgery, these methods offer reversible functional inhibition of specific neurons in the epileptogenic zone rather than irreversible destruction of brain tissue, and may therefore contribute to a future increase in treatment safety. Preclinical testing of new therapies is labor-intensive and bears a high risk of failure[36] (Table 12.2). Ultimate goals of personalized translational epileptology are to determine the individually contributing

Table 12.2 Novel approaches of treatment in epilepsy.

Novel approaches
Immunomodulatory interventions
MicroRNAs
DREADDs/optogenetics

DREADDs, Designer receptor exclusively activated by designer drugs.

patterns of genetic factors and various disease mechanisms in each single patient, and to tailor specific etiology-related therapies which take account of particular disease characteristics in the respective individual.

References

1. Shorvon SD, Andermann F, Guerrini R. *The Causes of Epilepsy*. Cambridge: Cambridge University Press; 2011.
2. Stafstrom CE. Pathophysiological mechanisms of seizures and epilepsy: a primer. In: Rho JM, Sankar R, Stafstrom CE, eds. *Epilepsy: Mechanisms, Models, and Translational Perspectives*. Boca Raton, FL: CRC; 2010:3–19.
3. Berg AT, Berkovic SF, Brodie MJ, et al. Revised terminology and concepts for organization of seizures and epilepsy: report of the ILAE Commission on Classification and Terminology, 2005–2009. *Epilepsia*. 2010;51:676–685.
4. Panebianco M, Rigby A, Weston J, Marson AG. Vagus nerve stimulation for partial seizures. *Cochrane Database Syst Rev*. 2015;2015(4):CD002896. Available from: https://doi.org/10.1002/14651858.CD002896.pub2. PMID: 25835947; PMCID: PMC7138043.
5. Mathern GW, Adelson PD, Cahan LD, Leite JP. Hippocampal neuron damage in human epilepsy: Meyer's hypothesis revisited. *Prog Brain Res*. 2002;135:237–251.
6. Coan AC, Cendes F. Epilepsy as progressive disorders: what is the evidence that can guide our clinical decisions and how can neuroimaging help? *Epilepsy Behav*. 2013;26:313–321.
7. Tejada J, Costa KM, Bertti P, Garcia-Cairasco N. The epilepsies: complex challenges needing complex solutions. *Epilepsy Behav*. 2013;26:212–228.
8. Stafstrom CE, Carmant L. Seizures and epilepsy: an overview for neuroscientists. *Cold Spring Harb Perspect Med*. 2015;5(6):a022426.
9. Britton JW, Frey LC, Hopp JL, et al. Introduction. In: St. Louis EK, Frey LC, eds. *Electroencephalography (EEG): An Introductory Text and Atlas of Normal and Abnormal Findings in Adults, Children, and Infants [Internet]*. Chicago, IL: American Epilepsy Society; 2016. Available from: https://www.ncbi.nlm.nih.gov/books/NBK390346/.
10. Gotman J. Computer analysis of EEG in epilepsy. In: Lopes Da Silva FH, Storm Van Leeuwn W, Remond A, eds. *Clinical application of Computer Analysis of EEG and Other Neurophysiological Signals. Handbook of Electroencephalography and Clinical Neurophysiology, Revised Series*. Vol. 2. Amsterdam: Elsevier; 1986:171–204.
11. Keller SS, Roberts N. Measurement of brain volume using MRI: software, techniques, choices and prerequisites. *J Anthropol Sci*. 2009;87:127–151. Erratum in *J Anthropol Sci*. 2010(88):251.
12. Lommers E, Simon J, Reuter G, et al. Multiparameter MRI quantification of microstructural tissue alterations in multiple sclerosis. *Neuroimage Clin*. 2019;23:101879. Available from: https://doi.org/10.1016/j.nicl.2019.101879.
13. Grunewald RA, Jackson GD, Connelly A, Duncan JS. MR detection of hippocampal disease in epilepsy: factors influencing T2 relaxation time. *AJNR Am J Neuroradiol*. 1994;15:1149–1156.
14. Burneo JG, Knowlton RC, Fought E, Martin R, Sawrie S, Kuzniecky RI. Chronic temporal lobe epilepsy: spatial extent and degree of metabolic dysfunction studied with magnetic resonance spectroscopy (MRS). *Epilepsy Res*. 2004;62:119–124.

15. Benedek K, Juhasz C, Chugani DC, Muzik O, Chuani HT. Longitudinal hangs in cortical glucose hypometabolism in children with intractable epilepsy. *J Child Neurol.* 2006;21 (1):26–31.
16. Spencer SS. The relative contributions of MRI, SPECT, and PET imaging in epilepsy. *Epilepsia.* 1994;35(Suppl.6):72–89.
17. Pearson TS, Akman C, Hinton VJ, Engelstad K, DeVivo DC. Phenotypic spectrum of glucose transporter type 1 deficiency syndrome (Glut1 DS). *Curr Neurol Neurosci Rep.* 2013;13:342.
18. Pearl PL, Taylor JL, Trzcinski S, Sokohl A. The pediatric neurotransmitter disorders. *J Child Neurol Neurosci Rep.* 2007;13:342.
19. Poduri A. When should genetic testing be performed in epilepsy patients? *Epilepsy Curr.* 2017;17(1):16–22. Available from: https://doi.org/10.5698/1535-7511-17.1.16.
20. Olson HE, Poduri A. Epilepsy: when to perform a genetic analysis. In: Miller JW, Goodkin HP, eds. *Epilepsy*. Chichester: Wiley; 2014:159–166.
21. Bui A, Kim H, Moroso M, Soltesz I. Microcircuits in epilepsy: heterogeneity and hub cells in network synchronization. *Cold Spring Harb Perspect Med.* 2015;5(11):a022855. Available from: https://doi.org/10.1101/cshperspect.a022855.
22. Neal EG, Chaffe H, Schwartz RH, et al. The ketogenic diet for the treatment of childhood epilepsy: a randomized controlled trial. *Lancet Neurol.* 2008;7(6):500–506.
23. Jl Sirven. Alternative therapies for seizures: promises and dangers. *Semin Neurol.* 2007;27:325–330.
24. Panebianco M, Sridharan K, Ramaratnam S. Yoga for epilepsy. *Cochrane Database Syst Rev.* 2017;10(10):CD001524. Available from: https://doi.org/10.1002/14651858.CD001524.pub3.
25. Arida RM, Guimaraes de Almeida AC, Cavlheiro EA, Scorza FA. Experimental and clinical findings from physical exercise as complementary therapy for epilepsy. *Epilepsy Behav.* 2013;26:273–278.
26. Reif PS, Tsai M-H, Helbig I, Rosenow F, Klein KM. Precision medicine in genetic epilepsies: break of dawn? *Expert Rev Neurother.* 2016;1–12.
27. Eriksson SH, Rugg-Gunn FJ, Symms MR, Barker GJ, Duncan JS. Diffusion tensor imaging in patients with epilepsy and malformations of cortical development. *Brain.* 2001;124(Pt3):617–626.
28. Golby AJ, Poldrack RA, Illes J, Chen D, Desmond JE, Gabrieli JD. Memory lateralization in medial temporal lobe epilepsy assessed by functional MRI. *Epilepsia.* 2002;43:855–863.
29. Dal-Col MLC, Terara-Bustamante VC, Valasco TR, Oliveira JAC, Sakamoto AC, Garcia-Cairasco N. Neuroethology application for the study of human temporal lobe epilepsy: from basic to applied sciences. *Epilepsy Behav.* 2009;14((Suppl.1):54–63.
30. Rosenow F, van Alphen N, Becker A, et al. Personalized translational epilepsy research – novel approaches and future perspectives. Part I: clinical and network analysis approaches. *Epilepsy & Behav.* 2017;76:13–18.
31. Iori V, Frigerio F, Vezzani A. Modulation of neuronal excitability by immune mediators in epilepsy. *Curr Opin Pharmacol.* 2016;26:118–123.
32. Bauer S, van Alphen N, Becker A, et al. Personalized translational epilepsy research – novel approaches and future perspectives. Part II: experimental and translational approaches. *Epilepsy & Behav.* 2017;76:7–12.
33. Rajman M, Metge F, Fiore R, et al. A microRNA-129–5p/Rbfox crosstalk coordinates homeostatic downscaling of excitatory synapses. *EMBO J.* 2017;36(12):1770–1787.

34. Woolf SH. The meaning of translational research and why it matters. *JAMA*. 2008;299 (2):211–213.
35. Tracy RP. Deep phenotyping": characterizing populations in the era of genomics and systems biology. *Curr Opin Lipidol*. 2008;19(2):151–157.
36. Katzel D, Nicholson E, Schorge S, Walker MC, Kullmann DM. Chemical-genetic attenuation of focal neocortical seizures. *Nat Commun*. 2014;5:3847.

Translational research in delayed cerebral ischemia

13

Vanitha Rajagopalan[1], *Vasudha Singhal*[2] *and Charu Mahajan*[1]
[1]Department of Neuroanesthesiology and Critical Care, All India Institute of Medical Sciences (AIIMS), New Delhi, India, [2]Department of Neuroanesthesiology and Critical Care Medanta-the Medicity, Gurgaon, India

Introduction

Aneurysmal subarachnoid hemorrhage (aSAH), which occurs due to rupture of an intracranial aneurysm, is a life-threatening condition and accounts for about 5% of all strokes.[1] Despite the advanced management modalities available and improved neurocritical care, ongoing preclinical and clinical research, the mortality and morbidity due to aSAH is unacceptably high.[2] Delayed cerebral ischemia (DCI) which is one of the most dreaded complications of aSAH, occurring in about 20% patients surviving the initial bleed, has been found to be a major culprit for this dismal outcome contributing to death, poor functional outcome, and prolonged length of intensive care unit (ICU) and hospital stays.

The limited success of recent clinical trials to provide an effective treatment and improve outcome has rekindled interest in research to develop reliable experimental models on the potential mechanisms underlying this condition. Animal models can provide the basis for dissecting the complex pathophysiology and screening potential medical therapies. Here comes in the role of translational research, which is a process of applying the findings of research conducted in the laboratory and in preclinical studies to developing trials and studies in humans and adopting and implementing the best of these practices to the community. Several experimental models of aSAH are available, but the exact extent to which DCI is recapitulated in each of these models is uncertain, thereby posing a major limitation in this regard. In this chapter, we discuss the role of translational research in DCI, the various models studied are briefly discussed including its utility and shortcomings and the future trends in this direction.

Experimental aneurysmal subarachnoid hemorrhage

The majority of data on the pathophysiological changes that occur at the time of aneurysm rupture is derived primarily from animal studies of SAH, mainly in rats. The animal models rely on iatrogenic damage to cerebral vessels to simulate

Perioperative Neuroscience. DOI: https://doi.org/10.1016/B978-0-323-91003-3.00002-7

aneurysm rupture and induce subarachnoid bleeding. While existing animal models mimic the injury, the lack of alignment of discrete measures of the underlying injury pathology between the animal models and human subjects poses difficulties in translating this to promising therapeutics. Hence, the applicability of rat models to the study of SAH in humans is questionable[3] and there is a need for new animal models of SAH with a high incidence of spontaneous cerebral aneurysm rupture.[4]

Characteristics of experimental SAH models

Most commonly studied species are rats (32%) and dogs (22%), followed by mice (17%), primates (16%), rabbits (12%), cats (1%), and pigs (1%).[5] SAH induction methods usually include blood injection into the prechiasmatic cistern, cisterna magna, or intrathecal space (61%), endovascular perforation of the internal carotid terminus or external perforation of the basilar artery (24%), direct clot placement around an intracranial artery (14%), or transection of a subarachnoid vein (1%). The most common endpoint studied in all the models was large artery vasospasm, particularly in larger animals. On the other hand, microvascular abnormalities are assessed in only a few recent studies, mostly in mice. Endpoint assessments were limited to the first week in most studies, although some continued for up to 48 days.

Perspectives on the current status of SAH animal models

There is a clear need for rigorous translational research using animal models to make further progress in the management of patients with SAH. The characteristics of an ideal animal model for SAH is described in Table 13.1. Although the human cerebral artery may be considered an ideal tool for studying the pathogenesis and treatment options of SAH, using live human brain vessels in vivo is not a viable option. The postmortem evaluation of arteries obtained from patients who died from SAH provides only minimal information.[6] Various animal species, including mice, rats, rabbits, dogs, primates, cats, pigs, and goats have been used to produce SAH models to closely mimic the physiological situation in humans.[7] Nonhuman primates are very similar in their genome, anatomy, and physiology to humans, and thus the models of nonhuman primates' SAH are believed to be the best candidates for replicating clinical SAH. The puncture monkey model through a small anterior

Table 13.1 Characteristics of the ideal animal model for subarachnoid hemorrhage.

Characteristics
Consistent deposition of blood in the subarachnoid space
Uniform degree of hemorrhage leading to increase in intracranial pressure
Close simulation of aneurysmal rupture by mechanism causing hemorrhage
Must be easily performed and at reasonable cost

craniotomy was first reported in 1968.[8] Nonhuman primates are usually time-consuming and demand complex surgical manipulation, such as anesthesia, angiography, and craniotomy, and are also not cost effective. Therefore the puncture technique was also adapted to small animals with the advancement of microsurgical technology. The rat is an excellent species, because of its relatively lower cost and ease of manipulation in the laboratory setting. Though technically more challenging, the endovascular puncture SAH model has also been produced in the mouse.[9] Recently, the therapeutic effect of potential drug molecules against early brain injury (EBI) has been examined on the mouse endovascular puncture model.[10] The advantage of a mouse SAH model would be the option to use transgenic mice, which is becoming the preferred research tool for gene-specific silencing in vivo. Using transgenic mice with spontaneous aSAH would better mimic the natural history seen in patients with SAH.

As there are no spontaneously occurring animal models of SAH, a number of animal models of SAH have been developed in various species to simulate SAH. SAH researchers can choose from a number of SAH animal models to suit their research objectives. The SAH models that mainly utilize two techniques: (1) endovascular punctures, in which puncture of an intracranial artery allows blood to quickly spread in the subarachnoid space; and (2) blood injection into the subarachnoid space after blood is obtained by the surgical exposure of a distant artery or vein.

Each animal model has its own advantages and disadvantages. It is generally accepted that the endovascular perforation model of SAH better mimics EBI, whereas the blood injection model better mimics the vasospasm that can occur after SAH.[11] In recent years, as the concept of EBI gains more popularity over delayed vasospasm, the blood injection model has been less favored as compared to the monofilament puncture model, since researchers are beginning to explore changes not only in the large cerebral arteries but also in the brain parenchyma.[12,13]

The endovascular puncture model was independently described in 1995 by Bederson et al.[12] and Veelken et al.[14] The surgical procedure involves perforation of the internal carotid bifurcation by a sharpened suture, which is inserted through the external carotid artery without craniotomy. As this method does not involve craniotomy, it best represents a clinical scenario mimicking the acute pathophysiological changes of an aneurysmal rupture in humans.[15,16] It results in considerably high rate of mortality, which ranges from 30% to 50% within 24 h after SAH.[17] However, the brief period of ischemia caused by application of an arterial clamp, the lack of control over the volume of the hemorrhage, and the high rate of morbidity and mortality significantly influences its use for studies exploring possible therapeutic options. Furthermore, the cerebral blood flow (CBF) on the side with the perforation tends to be lower than that of the nonperforated side, thus suggesting that this model results in some degree of laterality to the hemorrhagic event.

The blood injection model, in which blood is directly injected into the subarachnoid space either once or twice, is another widely used technique for inducing experimental SAH. This technique elicits early and delayed vasospasm in a variety of species, although its presence depends on the site of injection. More recently,

Marbacher et al. reported vasospasm in a number of SAH models: the most frequently injected volume of blood amount (mL), the peak onset of the vasospasm (day), and amount by which the blood vessel narrowed (%): mice endovascular puncture (a range, day 3, 20%–62%); rat single injection (0.3 mL, day 2, 19%–29%); rat double injection (0.3 mL, two times, day 7, 28%–47%); rabbit single injection (3 mL, day 3, 19%–55%); rabbit double injection (not established, day 5, not established); dog double injection (4–5 mL, two times, day 7, 45%–66%); and primate clot placement (5 mL, day 7, 32%–52%).[7] Recently, a novel technique of inducing SAH by extraintracranial blood shunt has been established to trigger delayed cerebral vasospasm under a controlled intracranial pressure (ICP).[18] In the future, experimental models of cerebral vasospasm should be improved so as to better mimic human SAH, in terms of having a direct hemorrhagic brain lesion under systolic pressure.[19]

Given the availability of various types of SAH models, it is imperative that the model chosen is appropriate for the objective of the study and that the model is replicable. In addition, genome-wide association studies are helpful to identify novel genetic factors that contribute to intracranial aneurysm susceptibility. Consequently, it is possible to produce an aneurysm on cerebral vessels experimentally using genetic technology.[20] Furthermore, ensuring reproducibility of results, the efficacy of a treatment option should be confirmed in multiple species and in multiple laboratories before beginning translation to a clinical study.[21]

Failure in current SAH studies

Despite the promising results seen in animal models, clinical studies have failed to translate the outcomes seen in animal models to human subjects.[22,23] The failure to translate animal studies could be due to methodological flaws during animal experiments. First, studies do not always indicate if the animals used in the experiments were randomized or how the randomization was performed. The studies may not be translatable if the animals used in the experiments were not randomly allocated. Furthermore, control groups are often inadequate, and some studies do not even include a control group. In addition, evaluations performed during experiments are not always blinded, which is essential for a nonbiased meaningful study. Second, most studies utilize young, nondiseased male animals, which therefore does not simulate the age, physical condition, or gender-specific risks seen in patients with SAH (who are more likely to be female, aged 45–55 years and have hypertension).[24–27] Furthermore, young rats are more resistant to the effect of the products of bilirubin oxidation compared to aged rats post-SAH.[28] Third, sample size is not always adequate for statistical analysis, which can lead to incorrect conclusions about the results of an experiment. Accurate power and statistical analyses are crucial for drawing valid conclusions from a study.

Further, in depth evaluation of outcomes in preclinical models is crucial to measure an agent's efficacy for the successful clinical translation of novel therapies.[29]

Mortality should be examined in experimental studies, as SAH is associated with very high mortality in patients. Neurological function also should be thoroughly evaluated, because this information is crucial for translation of animal studies. Therefore efficacy of therapies should be evaluated using tests to examine not only sensory motor function but also cognitive function, speech, and memory. Experimental studies need to incorporate an assessment of neurobehavioral outcomes in the acute phase as well as a more detailed evaluation of long-term neurological function.

It also is important to examine potential side effects of compounds being tested in animal studies in order to eliminate drugs that may be too toxic for clinical use. Using two or three gradual dosages is helpful to show dose dependent and potential toxic effects in experimental studies. The successful clinical translation of future studies requires the careful design of experiments, adequate controls, and rigorous testing in experimental SAH models. It may be prudent to develop a set of guidelines for SAH translational studies.[30]

Delayed brain injury

Delayed ischemic neurological deficit (DIND) is the leading cause of morbidity and mortality in patients who survive the initial impact of SAH and have had their aneurysm effectively treated.[31,32] Recent studies have demonstrated a host of critical, interrelated pathologies arising in the subacute phase of SAH as a result of EBI, and this phenomenon is designated as delayed brain injury (DBI). The advances in understanding of the pathophysiology of SAH have clarified that the mechanisms leading to EBI and DBI are not mutually exclusive and have many interrelated pathogenic triggers.

Delayed cerebral vasospasm: from extravascular blood to vessel lumen

Historically, cerebral vasospasm after SAH was considered to be a prolonged contraction of the smooth muscle cells and abnormal endothelial hypertrophy, inflammatory changes, and gene expression in the cerebral arteries, finally leading to tissue ischemia.[33] The central event in the contraction of vascular smooth muscles is an increase in the concentration of intracellular Ca^{2+}. Indeed, in the last century, there was a wide assumption that vasospasm was the main cause of poor outcomes for patients with SAH.[34] Several studies found a strong link between radiologically confirmed vasospasm and clinical symptoms of DCI.[32,35] As mentioned above, SAH-induced DCI is a clinical syndrome of focal neurological and cognitive disorder. Unpredictably, it occurs in 30% patients from 3 to 14 days after the initial bleeding.[36]

In vasospasm, overflowed blood from a ruptured aneurysm triggers a chain reaction of cerebral artery vasoconstriction, brain tissue infarction, and clinical condition deterioration. The most powerful predictors of vasospasm are the volume, density, and prolonged presence of subarachnoid blood surrounding the arteries.[37–40] Conversely, vasospasm was invariably absent in those patients with only a minimal subarachnoid blood load.[37] In the filament model of SAH, contraction of the large arteries in the circle of Willis in rats after SAH is ascribed to the persisting subarachnoid blood clot. Vasospasm began on the third day after the onset of SAH, and was maximal at 6–8 days, eventually lasting 2–3 weeks.[41] To sum up, cerebral vasospasm commonly takes place during DBI. However, there is not adequate evidence to conclude that vasospasm, in and of itself, could be used as a surrogate marker to monitor SAH progression and the efficacy of interventions.[42]

The effects of vasoactive agents have long been studied in regard to the pathogenesis of vasospasm.[43] These include neurogenic factors, biogenic amines (such as histamine and norepinephrine), 20-hydroxyeicosatetraenoic acid,[44] eicosanoids (such as prostaglandins, thromboxans, and leukotrienes), ion,[45] and free radicals.[46] Astrocytes and leukocytes release endothelin-1, a potent vasoconstrictor in response to SAH. In addition, endothelin receptors are upregulated.[47–50]

SAH increased the expression of G-protein-coupled vasoconstrictor receptors in cerebral arteries, including endothelin B receptor, 5-hydroxytryptamine receptor 1B, angiotensin II type 1 receptor, and thromboxane A2 receptors.[51,52] The Rho/Rho-kinase pathway is the main regulator of these actin-dependent cell functions in the pathogenesis of sustained smooth muscle cell contraction and vasospasm.[53,54] Therefore the combination of treatment with an inhibitor of RhoA, pitavastatin, and an inhibitor of Rho-kinase, fasudil, could extensively prevent cerebral vasospasm after SAH.[55] Furthermore, phosphorylation triggers SAH-induced vasculopathy in cerebral arteries, as determined by quantitative mass spectrometry, including focal adhesion complexes, extracellular regulated kinase 1/2, calcium calmodulin-dependent kinase II, STAT3, and c-Jun.[56] Tenascin-C, a matricellular protein, induces cerebral vasospasm via Toll-like receptor 4 and activation of Jun N-terminal kinases and p38 mitogen-activated protein kinases (MAPKs).[57,58] Animal experiments have shown a benefit from inhibiting the protein kinase C and MAPK pathways on cerebral vasoconstriction,[59,60] thus engendering enthusiasm that preconditioning stimuli may protect against SAH, as vasospasm-induced brain damage is anticipated after SAH.[61]

Microcirculatory spasm and dysfunction

Study of the changes in the microcirculation due to SAH started about 30 years ago.[62,63] The evolving concept of microvascular spasm has recently led to more attention being paid to the functional and structural changes in microcirculation.[64] Under normal conditions of cerebral autoregulation, the cerebral microvasculature can accommodate decreased perfusion pressure through vasodilation within the limits of autoregulation.[65] SAH results in failure of the microcirculation, including a significant decrease in the mass density of capillaries, vasocontraction in small arteries, and changes of the vessel wall that might lead to laminar, triangular, or round cortical infarcts.[66,67] Herz et al. demonstrated that topical applications of homologous blood alone could result in a 33% vasoconstriction in guinea pigs, which was consistent with a subsequent cortical infarction.[63] Progressive disturbances of the microcirculation result in, and/or contribute, to formation of microthrombi.[68] Disturbances in the microcirculation were found to be accompanied by disruption of the barriers of the intraparenchymal microvessels located distal or proximal to experimental clots in rats.[69] The thrombin receptor (protease-activated receptor-1) has been observed as playing a role in disruption of the endothelial barrier of hippocampus tissue.[70] In addition, a spasm may occur in microcirculation in dissociation with the state of extraparenchymal vessels.[71]

Apart from measuring CBF, which is the more common method, a study utilizing orthogonal polarizing spectral imaging showed that mono- and multisegmental cortical arterioles constrict after SAH with a reduction in diameter of up to 75%.[72] Ohkuma et al. noted that CBF was clearly associated with a prolonged circulation time in the cerebrum, even in patients without demonstrable spasms of large vessel, thus suggesting the impact of microcirculatory changes on constriction.[66] Although these data suggest that there are likely some microvascular perturbations after SAH, how these factors independently contribute to poor outcomes after SAH remains unknown.

Microthrombosis: an additional explanation for DCI

Microthrombosis has been attributed to the aggregation of platelets subsequent to the activation and amplification of the coagulation cascade after SAH. The presence of multiple microthrombi was, for the first time, confirmed in the autopsy of an SAH patient in 1983,[73] and have since then been confirmed in larger studies. A number of markers of microthrombosis are significantly upregulated after SAH, including fibrinopeptide A, tissue factor, and thrombin–antithrombin complexes.[74,75] Endothelial nitric oxide synthase knockout and decreased P-selectin reduce the formation of microthrombi after SAH.[68,76] Furthermore, the concentrations of these thrombotic markers were significantly elevated within days of SAH in patients who eventually developed DCI.[77]

It has been observed that approximately 30% of constricted arterioles were occluded by the microthrombi, suggesting that microthrombosis may be an etiology of the perfusion deficits and cerebral perfusion pressure-independent decrease in CBF after SAH.[78] Another line of investigation has suggested that microthrombi not only mechanically occlude the vessel, but also mediate vascular damage and constriction as the platelets aggregate and neutrophil infiltrates.[79,80] Intraluminal platelet aggregates can occur as soon as 10 min after SAH and are associated with injury to the vessel wall, including damage to the endothelium and the microvascular basal lamina.[81] Further, platelet aggregates can be amplified and stabilized through thrombin activation as well. The intracellular contents of platelets include serotonin, thrombin and adenosine diphosphate, and matrix metalloproteinase-9, which are known to potentially mediate vasoconstriction, degrade collagen IV in the microcirculation, and thus lead to disruption of the blood–brain barrier.[82,83]

Cortical spreading ischemia

Cortical spreading ischemia is a phenomenon in which ischemia is the consequence of neuronal/glial depolarization in experimental SAH and in patients with SAH.[84,85] Spreading ischemia in the cortex occurred when CBF reached 40%–70% compared to baseline, which was deemed to be 100%. Cortical spreading ischemia produced widespread cortical necrosis. In animal models, a number of pathologic conditions have been demonstrated to trigger cortical spreading depolarisation (CSD) with spreading ischemia, such as hypoxia, hypotension, transient ischemia from microemboli, low glucose level, nitric oxide depletion, and free hemoglobin and high extracellular K^+ from erythrocytes.[84,86,87]

Are we close to solving the mystery of SAH?

Recent data challenge the traditional concept of delayed vasospasm as the sole cause of DCI after SAH.[88,89] Beyond vasospasm, an emerging body of evidence at present suggests that DIND is likely to have a multifactorial etiology. Recently, in the randomized, double-blind, placebo-controlled, phase III trial, clazosentan (15 mg/h), an endothelin receptor antagonist, significantly decreased vasospasm-related morbidity, but did not significantly improved functional outcome in patients with aneurysmal SAH.[90,91] The mismatch between treatment of vasospasm and poor clinical outcome after SAH could have resulted from other underlying mechanisms of injury as opposed to vasospasm. To date, clinical trials have concentrated on vasospasm, which may be a wrong focus for SAH studies, since our increasing understanding of the SAH pathophysiology reveals EBI to be a major player in SAH-induced injury.[92]

A severe increase of ICP during the initial bleeding event has been identified as triggering subsequent pathophysiological changes after SAH. In recent years, EBI has evolved as a potential target to implement treatment modalities in patients with SAH, which could ameliorate some of the devastating injuries in the long term.[93] It can be suggested that vasospasm might be only a late manifestation of EBI,[94] because they share many of the same pathogenic factors. Further, EBI may also contribute to the pathogenesis of delayed neurological deficits. Nevertheless, research on EBI after SAH has been relatively limited; further studies are needed to clarify the exact mechanisms involved in EBI, which may reveal new therapeutic avenues that can be exploited in combination with antivasospasm medications.

Future directions

While there is no perfect standard treatment for this complex disease process, numerous attempts have been made to capture the complexity of clinical findings in DCI according to evidence-based criteria. Specificity is emphasized with respect to evidence-based clinically relevant outcomes at the primary level, which are particularly valuable in the research setting to potentially improve translation of research findings into clinical practice. The future direction involves determining the genetic predictors of DCI to study individual predisposition to cerebral vasoconstriction, CSD, and microthrombi formation. The uncertain pathophysiology of DCI has resulted in the lack of specific or sensitive serum or cerebrospinal fluid biomarker currently available for the early detection of this condition, thereby making identification of objective biomarkers for DCI a key research priority. Future studies into DCI may help to move forward both clinical care and research in this field.

Conclusion

In conclusion, DCI remains the major cause of mortality and morbidity in those patients who survive the initial bleed to hospital treatment. It now seems clear that DCI has a multifactorial etiology beyond pure cerebral vasoconstriction and that future management strategies are likely to be based on combination therapies targeting vasoconstriction, disturbed neuronal function, and disturbed neurovascular coupling.[85] The interval between aneurysmal rupture and the onset of DCI provides an optimal window for the application of preventative pharmacological therapies. Currently, however, the majority of therapeutic strategies in clinical use remain focused on improving cerebral perfusion in patients after the onset of DCI symptoms. More research is therefore required into the acute changes in cerebral physiology occurring in the first 72 h after SAH that may offer potential diagnostic and therapeutic targets for DCI in the future.

References

1. Sudlow CL, Warlow CP. Comparable studies of the incidence of stroke and its pathological types: results from an international collaboration. International Stroke Incidence Collaboration. *Stroke*. 1997;28(3):491–499.
2. Rabinstein AA, Lanzino G, Wijdicks EF. Multidisciplinary management and emerging therapeutic strategies in aneurysmal subarachnoid hemorrhage. *Lancet Neurol*. 2010;9 (5):504–519.

3. Swift DM, Solomon RA. Subarachnoid hemorrhage fails to produce vasculopathy or chronic blood flow changes in rats. *Stroke*. 1988;19(7):878–882.
4. Aoki T, Nishimura M. The development and the use of experimental animal models to study the underlying mechanisms of CA formation. *J Biomed Biotechnol*. 2011;2011: 1–10.
5. Oka F, Chung DY, Suzuki M, Ayata C. Delayed cerebral ischemia after subarachnoid hemorrhage: experimental-clinical disconnect and the unmet need. *Neurocrit Care*. 2020;32(1):238–251.
6. Crompton MR. The pathogenesis of cerebral infarction following the rupture of cerebral berry aneurysms. *Brain*. 1964;87:491–510.
7. Marbacher S, Fandino J, Kitchen ND. Standard intracranial in vivo animal models of delayed cerebral vasospasm. *Br J Neurosurg*. 2010;24(4):415–434.
8. Simeone FA, Ryan KG, Cotter JR. Prolonged experimental cerebral vasospasm. *J Neurosurg*. 1968;29(4):357–366.
9. Kamii H, Kato I, Kinouchi H, et al. Amelioration of vasospasm after subarachnoid hemorrhage in transgenic mice overexpressing CuZn-superoxide dismutase. *Stroke*. 1999;30(4):867–871.
10. Altay O, Hasegawa Y, Sherchan P, et al. Isoflurane delays the development of early brain injury after subarachnoid hemorrhage through sphingosine-related pathway activation in mice. *Crit Care Med*. 2012;40(6):1908–1913.
11. Lee JY, Sagher O, Keep R, Hua Y, Xi G. Comparison of experimental rat models of early brain injury after subarachnoid hemorrhage. *Neurosurgery*. 2009;65(2):331–343.
12. Bederson JB, Germano IM, Guarino L. Cortical blood flow and cerebral perfusion pressure in a new noncraniotomy model of subarachnoid hemorrhage in the rat. *Stroke*. 1995;26(6):1086–1091.
13. Titova E, Ostrowski RP, Zhang JH, Tang J. Experimental models of subarachnoid hemorrhage for studies of cerebral vasospasm. *Neurol Res*. 2009;31(6):568–581.
14. Veelken JA, Laing RJ, Jakubowski J. The Sheffield model of subarachnoid hemorrhage in rats. *Stroke*. 1995;26(7):1279–1283.
15. Schwartz AY, Masago A, Sehba FA, Bederson JB. Experimental models of subarachnoid hemorrhage in the rat: a refinement of the endovascular filament model. *J Neurosci Methods*. 2000;96(2):161–167.
16. Schwartz AY, Sehba FA, Bederson JB. Decreased nitric oxide availability contributes to acute cerebral ischemia after subarachnoid hemorrhage. *Neurosurgery*. 2000;47(1): 208–214.
17. Prunell GF, Mathiesen T, Svendgaard NA. Experimental subarachnoid hemorrhage: cerebral blood flow and brain metabolism during the acute phase in three different models in the rat. *Neurosurgery*. 2004;54(2):426–436.
18. Marbacher S, Sherif C, Neuschmelting V, et al. Extra-intracranial blood shunt mimicking aneurysm rupture: intracranial-pressure-controlled rabbit subarachnoid hemorrhage model. *J Neurosci Methods*. 2010;191(2):227–233.
19. Pluta RM, Hansen-Schwartz J, Dreier J, et al. Cerebral vasospasm following subarachnoid hemorrhage: time for a new world of thought. *Neurol Res*. 2009;31(2): 151–158.
20. Suda N, Moriyama K, Ganburged G. Effect of angiotensin II receptor blocker on experimental periodontitis in a mouse model of Marfan syndrome. *Infect Immun*. 2013;81(1): 182–188.
21. Tajiri N, Dailey T, Metcalf C, et al. In vivo animal stroke models: a rationale for rodent and non-human primate models. *Transl Stroke Res*. 2013;4(3):308–321.

22. Feuerstein GZ, Zaleska MM, Krams M, et al. Missing steps in the STAIR case: a translational medicine perspective on the development of NXY-059 for treatment of acute ischemic stroke. *J Cereb Blood Flow Metab*. 2008;28(1):217–219.
23. Savitz SI. A critical appraisal of the NXY-059 neuroprotection studies for acute stroke: a need for more rigorous testing of neuroprotective agents in animal models of stroke. *Exp Neurol*. 2007;205(1):20–25.
24. Kassell NF, Haley Jr EC, Apperson-Hansen C, Alves WM. Randomized, double-blind, vehicle-controlled trial of tirilazad mesylate in patients with aneurysmal subarachnoid hemorrhage: a cooperative study in Europe, Australia, and New Zealand. *J Neurosurg*. 1996;84(2):221–228.
25. Kongable GL, Lanzino G, Germanson TP, et al. Gender-related differences in aneurysmal subarachnoid hemorrhage. *J Neurosurg*. 1996;84(1):43–48.
26. Lanzino G, Kassell NF. Double-blind, randomized, vehicle-controlled study of high-dose tirilazad mesylate in women with aneurysmal subarachnoid hemorrhage. Part II. A cooperative study in North America. *J Neurosurg*. 1999;90(6):1018–1024.
27. Lanzino G, Kassell NF, Dorsch NW, et al. Double-blind, randomized, vehicle-controlled study of high-dose tirilazad mesylate in women with aneurysmal subarachnoid hemorrhage. Part I. A cooperative study in Europe, Australia, New Zealand, and South Africa. *J Neurosurg*. 1999;90(6):1011–1017.
28. Clark JF, Harm A, Saffire A, Biehle SJ, Lu A, Pyne-Geithman GJ. Bilirubin oxidation products seen post subarachnoid hemorrhage have greater effects on aged rat brain compared to young. *Acta Neurochir Suppl*. 2011;110(Pt 1):157–162.
29. Knight Jr DR, Smith AH, Schroeder RL, et al. Effects of age on noninvasive assessments of vascular function in nonhuman primates: implications for translational drug discovery. *J Transl Med*. 2013;11:101.
30. Lapchak PA, Zhang JH, Noble-Haeusslein LJ. RIGOR guidelines: escalating STAIR and STEPS for effective translational research. *Transl Stroke Res*. 2013;4:279–285.
31. Crowley RW, Medel R, Kassell NF, Dumont AS. New insights into the causes and therapy of cerebral vasospasm following subarachnoid hemorrhage. *Drug Discov Today*. 2008;13:254–260.
32. Fergusen S, Macdonald RL. Predictors of cerebral infarction in patients with aneurysmal subarachnoid hemorrhage. *Neurosurgery*. 2007;60:658–667.
33. Rothoerl RD, Ringel F. Molecular mechanisms of cerebral vasospasm following aneurysmal SAH. *Neurol Res*. 2007;29:636–642.
34. Kassell NF, Torner JC, Haley Jr EC, Jane JA, Adams HP, Kongable GL. The international cooperative study on the timing of aneurysm surgery. Part 1: overall management results. *J Neurosurg*. 1990;73:18–36.
35. Rabinstein AA, Friedman JA, Weigand SD, et al. Predictors of cerebral infarction in aneurysmal subarachnoid hemorrhage. *Stroke*. 2004;35:1862–1866.
36. Dorsch NW, King MT. A review of cerebral vasospasm in aneurysmal subarachnoid haemorrhage Part I: incidence and effects. *J Clin Neurosci*. 1994;1:19–26.
37. Fisher CM, Kistler JP, Davis JM. Relation of cerebral vasospasm to subarachnoid hemorrhage visualized by computerized tomographic scanning. *Neurosurgery*. 1980;6:1–9.
38. Loch Macdonald R. Management of cerebral vasospasm. *Neurosurg Rev*. 2006;29:179–193.
39. Reilly C, Amidei C, Tolentino J, Jahromi BS, Macdonald RL. Clot volume and clearance rate as independent predictors of vasospasm after aneurysmal subarachnoid hemorrhage. *J Neurosurg*. 2004;101:255–261.

40. Suzuki J, Komatsu S, Sato T, Sakurai Y. Correlation between CT findings and subsequent development of cerebral infarction due to vasospasm in subarachnoid haemorrhage. *Acta Neurochir (Wien)*. 1980;55:63–70.
41. Wilkins RH. Cerebral vasospasm. *Crit Rev Neurobiol*. 1990;6:51–77.
42. Nolan CP, Macdonald RL. Can angiographic vasospasm be used as a surrogate marker in evaluating therapeutic interventions for cerebral vasospasm? *Neurosurg Focus*. 2006;21:E1.
43. Roman RJ, Renic M, Dunn KM, Takeuchi K, Hacein-Bey L. Evidence that 20-HETE contributes to the development of acute and delayed cerebral vasospasm. *Neurol Res*. 2006;28:738–749.
44. Mulligan SJ, MacVicar BA. Calcium transients in astrocyte endfeet cause cerebrovascular constrictions. *Nature*. 2004;431:195–199.
45. Girouard H, Iadecola C. Neurovascular coupling in the normal brain and in hypertension, stroke, and Alzheimer disease. *J Appl Physiol*. 2006;100:328–335.
46. Nishizawa S, Laher I. Signaling mechanisms in cerebral vasospasm. *Trends Cardiovasc Med*. 2005;15:24–34.
47. Beg SA, Hansen-Schwartz JA, Vikman PJ, Xu CB, Edvinsson LI. ERK1/2 inhibition attenuates cerebral blood flow reduction and abolishes ET(B) and 5-HT(1B) receptor upregulation after subarachnoid hemorrhage in rat. *J Cereb Blood Flow Metab*. 2006;26:846–856.
48. Fassbender K, Hodapp B, Rossol S, et al. Endothelin-1 in subarachnoid hemorrhage: an acute-phase reactant produced by cerebrospinal fluid leukocytes. *Stroke*. 2000;31: 2971–2975.
49. Hansen-Schwartz J, Hoel NL, Zhou M, Xu CB, Svendgaard NA, Edvinsson L. Subarachnoid hemorrhage enhances endothelin receptor expression and function in rat cerebral arteries. *Neurosurgery*. 2003;52:1188–1194.
50. Pluta RM, Boock RJ, Afshar JK, et al. Source and cause of endothelin-1 release into cerebrospinal fluid after subarachnoid hemorrhage. *J Neurosurg*. 1997;87:287–293.
51. Hansen-Schwartz J, Hoel NL, Xu CB, Svendgaard NA, Edvinsson L. Subarachnoid hemorrhage-induced upregulation of the 5-HT1B receptor in cerebral arteries in rats. *J Neurosurg*. 2003;99:115–120.
52. Povlsen GK, Waldsee R, Ahnstedt H, Kristiansen KA, Johansen FF, Edvinsson L. In vivo experimental stroke and in vitro organ culture induce similar changes in vasoconstrictor receptors and intracellular calcium handling in rat cerebral arteries. *Exp Brain Res*. 2012;219:507–520.
53. Satoh S, Takayasu M, Kawasaki K, et al. Antivasospastic effects of hydroxyfasudil, a Rho-kinase inhibitor, after subarachnoid hemorrhage. *J Pharmacol Sci*. 2012;118: 92–98.
54. Takai Y, Sasaki T, Tanaka K, Nakanishi H. Rho as a regulator of the cytoskeleton. *Trends Biochem Sci*. 1995;20:227–231.
55. Naraoka M, Munakata A, Matsuda N, Shimamura N, Ohkuma H. Suppression of the Rho/Rho-kinase pathway and prevention of cerebral vasospasm by combination treatment with statin and fasudil after subarachnoid hemorrhage in rabbit. *Transl Stroke Res*. 2013;4:368–374.
56. Parker BL, Larsen MR, Edvinsson LI, Povlsen GK. Signal transduction in cerebral arteries after subarachnoid hemorrhage-a phosphoproteomic approach. *J Cereb Blood Flow Metab*. 2013 Aug;33(8):1259–1269.
57. Fujimoto M, Suzuki H, Shiba M, et al. Tenascin-C induces prolonged constriction of cerebral arteries in rats. *Neurobiol Dis*. 2013;55:104–109.

58. Suzuki H, Shiba M, Fujimoto M, et al. Matricellular protein: a new player in cerebral vasospasm following subarachnoid hemorrhage. *Acta Neurochir Suppl*. 2013;115:213–218.
59. Beg SS, Hansen-Schwartz JA, Vikman PJ, Xu CB, Edvinsson LI. Protein kinase C inhibition prevents upregulation of vascular ET(B) and 5-HT(1B) receptors and reverses cerebral blood flow reduction after subarachnoid haemorrhage in rats. *J Cereb Blood Flow Metab*. 2007;27:21–32.
60. Hansen-Schwartz J, Ansar S, Edvinsson L. Cerebral vasoconstriction after subarachnoid hemorrhage–role of changes in vascular receptor phenotype. *Front Biosci*. 2008;13:2160–2164.
61. Wang MM, Xi G, Keep RF. Should the STAIR criteria be modified for preconditioning studies? *Transl Stroke Res*. 2013;4:3–14.
62. Wiernsperger N, Schulz U, Gygax P. Physiological and morphometric analysis of the microcirculation of the cerebral cortex under acute vasospasm. *Stroke*. 1981;12: 624–627.
63. Herz DA, Baez S, Shulman K. Pial microcirculation in subarachnoid hemorrhage. *Stroke*. 1975;6:417–424.
64. Kozniewska E, Michalik R, Rafalowska J, et al. Mechanisms of vascular dysfunction after subarachnoid hemorrhage. *J Physiol Pharmacol*. 2006;57(Suppl 11):145–160.
65. Paulson OB, Strandgaard S, Edvinsson L. Cerebral autoregulation. *Cerebrovasc Brain Metab Rev*. 1990;2:161–192.
66. Ohkuma H, Manabe H, Tanaka M, Suzuki S. Impact of cerebral microcirculatory changes on cerebral blood flow during cerebral vasospasm after aneurysmal subarachnoid hemorrhage. *Stroke*. 2000;31:1621–1627.
67. Weidauer S, Vatter H, Beck J, et al. Focal laminar cortical infarcts following aneurysmal subarachnoid haemorrhage. *Neuroradiology*. 2008;50:1–8.
68. Sabri M, Ai J, Lakovic K, D'Abbondanza J, Ilodigwe D, Macdonald RL. Mechanisms of microthrombi formation after experimental subarachnoid hemorrhage. *Neuroscience*. 2012;224:26–37.
69. Johshita H, Kassell NF, Sasaki T. Blood-brain barrier disturbance following subarachnoid hemorrhage in rabbits. *Stroke*. 1990;21:1051–1058.
70. Yan J, Manaenko A, Chen S, et al. Role of SCH79797 in maintaining vascular integrity in rat model of subarachnoid hemorrhage. *Stroke*. 2013;44:1410–1417.
71. Ohkuma H, Suzuki S. Histological dissociation between intra- and extraparenchymal portion of perforating small arteries after experimental subarachnoid hemorrhage in dogs. *Acta Neuropathol*. 1999;98:374–382.
72. Uhl E, Lehmberg J, Steiger HJ, Messmer K. Intraoperative detection of early microvasospasm in patients with subarachnoid hemorrhage by using orthogonal polarization spectral imaging. *Neurosurgery*. 2003;52:1307–1315.
73. Suzuki S, Suzuki M, Iwabuchi T, Kamata Y. Role of multiple cerebral microthrombosis in symptomatic cerebral vasospasm: with a case report. *Neurosurgery*. 1983;13: 199–203.
74. Stein SC, Browne KD, Chen XH, Smith DH, Graham DI. Thromboembolism and delayed cerebral ischemia after subarachnoid hemorrhage: an autopsy study. *Neurosurgery*. 2006;59:781–787.
75. Vergouwen MD, Vermeulen M, Coert BA, Stroes ES, Roos YB. Microthrombosis after aneurysmal subarachnoid hemorrhage: an additional explanation for delayed cerebral ischemia. *J Cereb Blood Flow Metab*. 2008;28:1761–1770.
76. Sabri M, Lass E, Macdonald RL. Early brain injury: a common mechanism in subarachnoid hemorrhage and global cerebral ischemia. *Stroke Res Treat*. 2013;2013:394036.

77. Dorsch N. A clinical review of cerebral vasospasm and delayed ischaemia following aneurysm rupture. *Acta Neurochir Suppl*. 2011;110:5–6.
78. Friedrich B, Muller F, Feiler S, Scholler K, Plesnila N. Experimental subarachnoid hemorrhage causes early and long-lasting microarterial constriction and microthrombosis: an in-vivo microscopy study. *J Cereb Blood Flow Metab*. 2012;32:447–455.
79. Friedrich V, Flores R, Muller A, Bi W, Peerschke EI, Sehba FA. Reduction of neutrophil activity decreases early microvascular injury after subarachnoid haemorrhage. *J Neuroinflammation*. 2011;8:103.
80. Sehba FA, Mostafa G, Friedrich Jr V, Bederson JB. Acute microvascular platelet aggregation after subarachnoid hemorrhage. *J Neurosurg*. 2005;102:1094–1100.
81. Friedrich V, Flores R, Muller A, Sehba FA. Escape of intraluminal platelets into brain parenchyma after subarachnoid hemorrhage. *Neuroscience*. 2010;165:968–975.
82. Scholler K, Trinkl A, Klopotowski M, et al. Characterization of microvascular basal lamina damage and blood-brain barrier dysfunction following subarachnoid hemorrhage in rats. *Brain Res*. 2007;1142:237–246.
83. Sehba FA, Mostafa G, Knopman J, Friedrich Jr V, Bederson JB. Acute alterations in microvascular basal lamina after subarachnoid hemorrhage. *J Neurosurg*. 2004;101: 633–640.
84. Dreier JP, Korner K, Ebert N, et al. Nitric oxide scavenging by hemoglobin or nitric oxide synthase inhibition by N-nitro-L-arginine induces cortical spreading ischemia when K+ is increased in the subarachnoid space. *J Cereb Blood Flow Metab*. 1998;18:978–990.
85. Dreier JP, Major S, Manning A, et al. Cortical spreading ischaemia is a novel process involved in ischaemic damage in patients with aneurysmal subarachnoid haemorrhage. *Brain*. 2009;132:1866–1881.
86. Dreier JP, Ebert N, Priller J, et al. Products of hemolysis in the subarachnoid space inducing spreading ischemia in the cortex and focal necrosis in rats: a model for delayed ischemic neurological deficits after subarachnoid hemorrhage? *J Neurosurg*. 2000;93: 658–666.
87. Nozari A, Dilekoz E, Sukhotinsky I, et al. Microemboli may link spreading depression, migraine aura, and patent foramen ovale. *Ann Neurol*. 2010;67:221–229.
88. Kusaka G, Ishikawa M, Nanda A, Granger DN, Zhang JH. Signaling pathways for early brain injury after subarachnoid hemorrhage. *J Cereb Blood Flow Metab*. 2004;24: 916–925.
89. Macdonald RL, Pluta RM, Zhang JH. Cerebral vasospasm after subarachnoid hemorrhage: the emerging revolution. *Nat Clin Pract Neurol*. 2007;3:256–263.
90. Macdonald RL, Higashida RT, Keller E, et al. Randomized trial of clazosentan in patients with aneurysmal subarachnoid hemorrhage undergoing endovascular coiling. *Stroke*. 2012;43:1463–1469.
91. Sabri M, Ai J, Macdonald RL. Dissociation of vasospasm and secondary effects of experimental subarachnoid hemorrhage by clazosentan. *Stroke*. 2011;42:1454–1460.
92. Gomis P, Graftieaux JP, Sercombe R, Hettler D, Scherpereel B, Rousseaux P. Randomized, double-blind, placebo-controlled, pilot trial of high-dose methylprednisolone in aneurysmal subarachnoid hemorrhage. *J Neurosurg*. 2010;112:681–688.
93. Cahill J, Calvert JW, Zhang JH. Mechanisms of early brain injury after subarachnoid hemorrhage. *J Cereb Blood Flow Metab*. 2006;26:1341–1353.
94. Povlsen GK, Johansson SE, Larsen CC, Samraj AK, Edvinsson L. Early events triggering delayed vasoconstrictor receptor upregulation and cerebral ischemia after subarachnoid hemorrhage. *BMC Neurosci*. 2013;14:34.

Translation research in pain medicine

14

Ryan R. Budwany[1], *Yeshvant A. Navalgund*[1] *and Alaa Abd-Elsayed*[2]
[1]Center for Integrative Pain Management, West Virginia University, Morgantown, WV, United States, [2]Pain Management Services, UW Health, Madison, WI, United States

Translational research

Translational research is the process of applying knowledge from basic science research and clinical trials to techniques and tools that address critical medical needs. The term is often used interchangeably with translational medicine or translational science.

Additionally, it may be referred to as "bench-to-bedside". However, translational research should also work in the reverse direction of bedside to bench, allowing the flow of insight from clinicians to shape the next generation of basic science and clinical research. In this regard, translational pain research strives to push both pain research and pain medicine forward. Investigators can expect to see funding for translational research growing at a dizzying pace over the next several decades.[1]

There are many phases of the translational research and some include nonmedical tasks (information technology, engineering, nanotechnology, biochemistry, animal research, economy, sociology, psychology, politics, etc.).[2] Fig. 14.1 provides a

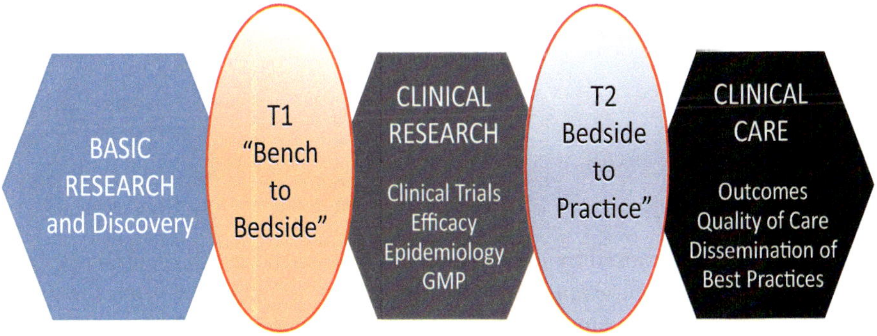

Figure 14.1 Schematic transitions from the "Bench to Bedside" (T1) and "Bedside to Practice" (T2) translational medical research and development model. The process of invention and then development is bidirectionally promoted by academic inquiry, as well as astute clinical observations. "Bench to Bedside" is catalyzed by first principles inquiry and the "Bedside to Bench" direction is catalyzed by a clinical observation and an articulated "if only I had . . .." to solve this clinical problem or fill this unmet need in my practice of medicine.

Perioperative Neuroscience. DOI: https://doi.org/10.1016/B978-0-323-91003-3.00001-5

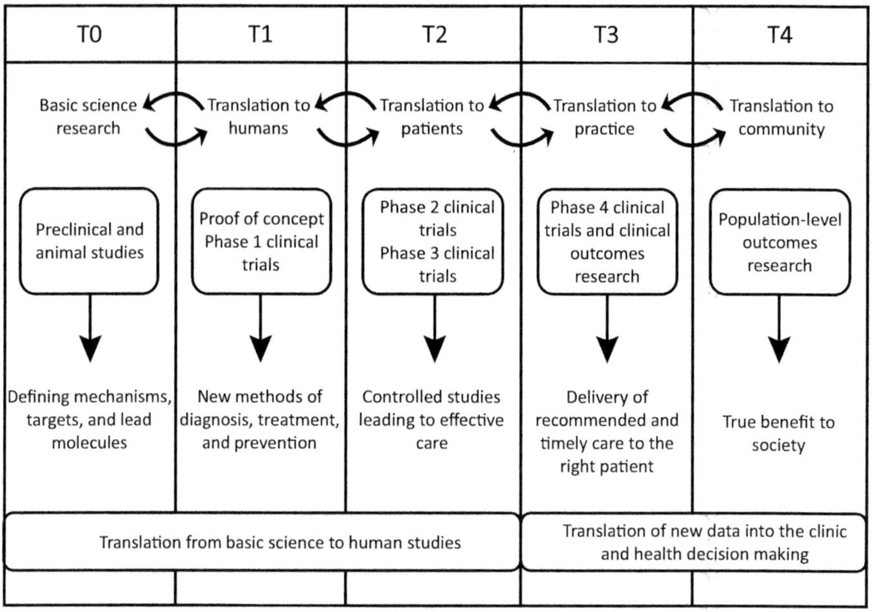

Figure 14.2 Operational phases of translational research.

simple view of the translational research process while Fig. 14.2 provides more granularity on the operational phases.[3]

Between each phase is a gap, or roadblock, that is characterized by the lack of funding or support needed for progression to the next phase. The most significant gap is between T0 and T1; it is called the *valley of death* because most fundamental research studies do not go beyond it and therefore do not enter the clinical phases.[2]

Types of pain

Neuropathic pain

Neuropathic pain is caused by injury to nerve tissue, including the central or peripheral nervous system and even the autonomic system. It originates from the pathology of the nervous system and is characterized by abnormal hypersensitivity to stimuli (hyperalgesia) and nociceptive responses to nonnoxious stimuli (allodynia). The conditions and the pathophysiological states that determine the onset of neuropathic pain are heterogeneous, such as metabolic disorders, neuropathy caused by viral infections, and autoimmune diseases affecting the central nervous system.[4] A best estimate of population prevalence of pain with neuropathic characteristics is likely to lie between 6.9% and 10%.[5]

Neuropathic pain is not a single disease, but instead a syndrome that can be caused by a number of diverse etiologies, and its clinical manifestations vary significantly.[6] As a result, neuropathic pain is widely recognized as one of the most difficult pain syndromes to manage, and outcomes often are unsatisfactory.

Nociceptive pain

Nociceptive pain is caused by the activation of nociceptive nerve fibers by physical tissue destruction or by chemical, pressure, or thermal processes. Nociceptive pain is usually acute, develops in response to a specific injury or situation, and tends to go away as the affected area heals. Nociceptive pain typically responds well to medicinal therapy, including both over-the-counter and prescription medications.

Models in pain management

There are currently three primary models in approaching pain management.

Medical management

The medical management approach involves using medicines to treat pain. While nociceptive pain is often effectively treated with medications such as NSAIDs and opioids, medical management of neuropathic pain is more complex. The Special Interest Group on Neuropathic Pain (NeuPSIG) proposed gabapentinoids, tricyclic antidepressants, and selective serotonin–norepinephrine reuptake inhibitors as the first-line drugs for neuropathic pain. Lidocaine, Capsaicin, and Tramadol have been proposed as the second-line treatment, while strong opioids (Morphine and Oxycodone) and botulinum toxin-A were included as third-line treatments for peripheral neuropathic pain.[7]

Interventional management

The interventional approach to managing neuropathic pain that is refractory to pharmacologic management includes steroid injection, neural blockades, epidural injections, implantable neuromodulation systems including intrathecal drug pumps and spinal cord stimulators, and neurosurgical procedures.

Intrathecal therapy is a viable and relatively safe option for the treatment of cancer- and noncancer-related pain.[8] Intrathecal therapy is often implemented when appropriate titration of a pharmacologic agent cannot be achieved due to significant side-effects, as well as when adequate pain relief cannot be achieved despite high doses of opioids. Currently morphine, ziconotide, and baclofen are the only medications approved by the US Food and Drug Administration (FDA) for intrathecal use.

Spinal cord stimulation (SCS) has become an important tool in the management of intractable, noncancer-related, neuropathic pain and is relatively safe because of its minimally invasive and reversible characteristics.[9] Originally based on the proposed concept of the gate control theory of pain, SCS has advanced quite rapidly in recent years. Mechanisms of action beyond gate control theory, such as the direct modulation of wide–dynamic range neurons, the modulation of activity in cortical and subcortical brain regions, and activation of descending inhibitory pathways, have been proposed.[10] Randomized controlled studies support the efficacy of SCS in treating failed back surgery syndrome and complex regional pain syndrome. Similar studies of neurostimulation for peripheral neuropathic pain, postamputation pain, postherpetic neuralgia, and other causes of nerve injury are needed.[9]

Today's neuromodulation systems are no longer an extension of anecdotal data driven by industry-funded science, but evidence-based treatment for chronic pain patients that demonstrate a significant improvement in efficacy compared to alternative treatment modalities. Historically, patients living with chronic pain have had to undergo significant trials of opioid medications, other pharmacological therapies, and other procedural interventions in order to "qualify" for neuromodulation therapy, which was positioned as a last-line therapy. However, studies have shown that patients who received high doses of opioids prior to SCS were at risk for poor outcomes including a higher risk for system explant by 57%, longer postoperative stays, and dampened improvement of pain scores.[11]

The future of interventional pain management includes requiring a comprehensive assessment of chronic pain patients and more rapid progression through various modalities with avoidance or extremely limited use of opioid therapy.[11]

Integrative management

Integrative pain management moves away from the reliance on medical management and applies a more holistic approach that targets both the mind and body to treat pain. An integrative approach may address mental health issues with individual or group therapy and incorporate a number of different modalities including movement therapy, relaxation therapy, transcutaneous electrical nerve stimulation, occupational therapy, physical therapy, dietary modification, spinal manipulation, acupuncture, Tai Chi/Qigong, yoga, massage, mindfulness and meditation.

Before the advent of modern medical practices and interventions for pain control, pain was most frequently managed with an integrative approach. Returning to the integrative approach is appealing in a number of ways. First, due to concerns of tolerance, dependence, and addiction in light of the ensuing opioid crisis, both patients and clinicians are more often seeking alternatives to medical management. In addition, an integrative approach is often a collaboration between patient and physician resulting in a form of personalized medicine that appeals to many patients today. Today's chronic pain patients have typically spent considerable time researching a wide range of pain management options on the internet and are open to new ideas. It is an ideal scenario for a pain physician to help guide them through the process of undergoing various treatments while balancing their desire for additional pain relief with their level of comfort with more invasive or aggressive treatments.

While the integrative model appears to be the model that pain management will concentrate on in the next decades, to date there have been only a few studies published that have focused on how patients do when going through integrative programs. Integrative pain management should be an important area of translational research in the years to come.

Current trends in care

The fact that each patient's pain experience is unique and a "one size fits all" approach has not been successful, in combination with rapid development of new personalized technology ("wearable devices") that provide rich, diverse sources of digital data and the ability to develop and continually refine models through analysis of large amounts of data and machine learning, makes personalized medicine a natural trend for pain medicine. Pain research and pain medicine have always been hindered by the lack of an objective pain measurement. The ability to collect and analyze data like never before is allowing pain medicine to realize the ability to utilize data from many sources to create predictive models and apply those models to both individuals and populations.

These sources may include the following:

- *Physiological data*: heart rate and heart rate variability, eye fixations and pupil diameter, galvanic skin response, emotion tracking through artificial intelligence analysis of facial actions, and thermography.
- *Cognitive data*: response inhibition, attention, and working memory evaluated via app assessments
- *Biomarkers*: stress, anxiety, and inflammation assessed via blood, sweat, and saliva samples (biomarkers cortisol, NPY, IL6, and TNFα)
- *Functional/behavioral*: movement, gate and posture, sleep, and psychosocial assessments.

These data are beginning to be collected, stored, and analyzed in patient clouds to create predictive models. As these data accumulate, the models will be come more accurate at improving health and wellness, suggesting lifestyle changes, preventing injuries, improving performance, predicting longevity of career and function, detecting disease, and suggesting treatments and interventions such as neuromodulation devices if benefits are predicted.

This type of data-driven treatment management throughout the pain care continuum will guide outcomes and result in higher patient engagement & satisfaction.

Another exciting development in pain medicine is Harvard MedTech's use of virtual reality to retrain the brain in treating chronic pain. This noninvasive, nonpharmacologic therapy is thought to reprogram the brain's processing and perception of, and response, to pain stimuli through immersive visual and auditory experiences, and thereby realigning neural pathways. The therapy helps patients reduce pain, relieve stress, and build resilience.[12]

Immense opportunity for translational research in pain medicine

Despite huge scientific and technological advances, and tremendously increased research and development costs, drugs are more likely to fail in clinical development today than they were in the 1970s. The credibility of efficacy data obtained from animal disease models has lately been called into question, in biomedical research in general and in the pain field specifically. For pain studies, the likelihood of eventual FDA approval of a drug entering phase I studies has been reported at 10.7%.[13]

Some have blamed animal models for these translational difficulties, and perhaps rightfully so. Basic research on pain and pain modulation is unique among many branches of biomedical research in that objective measures such as nociceptive responses to thermal or mechanical stimulation in experimental animals are used to investigate only the sensory discriminative aspect of the subjective pain sensation.[14]

While translational research is significant for all areas of medicine, its implications for pain medicine are essential for pushing research into a new era where models are more multidimensional and representative of the way pain is experienced by patients, such as in the presence of psychosocial comorbidities of anxiety and depression. The opportunity is ripe for translational research to guide pain research and pain medicine in several important areas.

Bridging the gap between pain research and clinical pain management

Prior to the coronavirus pandemic, the opioid crisis was likely the most profound public health crisis the United States had faced. Alarmingly, the overwhelming majority of opioid abusers begin their addiction with prescription medications, primarily for chronic pain.[15] While the opioid crisis has certainly resulted in changes in the way these medications are prescribed, both from a regulatory and ethical perspective, the move away from the medical model is also partially driven by the fact that this model was not working effectively in treating pain. The interventional model has provided pain relief to many patients, but the interventional treatments have risks, significant cost, and some level of burden to patients for things like intrathecal pump medication refills, dosing adjustments, and battery replacements. Spinal cord stimulators also require battery replacements and occasional setting adjustments. These shortcomings in both the medical and interventional models again point towards the integrative model being the future of pain management. The integrative model can incorporate modalities of the other models when appropriate but can focus on a number of the alternatives that don't carry the risks or costs that medications and interventions do. For example, an integrative approach might incorporate pain medications for a short time with the goal of using yoga and acupuncture in a way that eventually allows the patient to wean off the medication and manage their pain solely with the alternative modalities.

Examples of translational pain medicine

Ziconotide

Therapeutic development of the drug ziconotide (Prialt) provides an example of translational pain medicine that was primarily a bench-to-bedside process. Ziconotide is derived from the venom of a giant marine snail called Conus magus and is 1000 times as powerful as morphine.

Ziconotide development proceeded through the translational research process as described below:

Phase T0: Basic science research demonstrated in laboratory rats that intravenous ziconotide degraded rapidly, produced no detectable intermediate, and cleared quickly from both the cerebrospinal fluid and the circulatory system. Antinociceptive effects were also demonstrated in various animal models of acute and chronic pain, and toxicology studies supported its use, even at higher doses than would be used in humans.

Phase T1: Translation to humans started with an open-label feasibility study conducted in 31 patients. Preliminary analgesic evidence was positive; safety data indicated adverse effects diminished or resolved when doses were decreased or ceased. These results generated promise in a new treatment that could be translated to patients in a controlled trial.

Phase T2: Translation to patients resulted in the first ziconotide RCT in 1996 in patients with pain from cancer or AIDS.[16] Analgesic results were positive, but there were a large amount and variety of adverse effects. The next RCT included an expanded patient population (severe chronic nonmalignant pain unresponsive to conventional therapy). This trial was similar in that it demonstrated positive analgesic results, but a high incidence of ziconotide-related adverse effects due to the rapid titration and high doses administered.[17] A third RCT followed and confirmed that slow titration to a low maximum dose resulted in significant improvement in pain and was better tolerated than in the two previous trials.[18]

Phase T3/T4: Upon FDA approval in 2004, ziconotide research was able to translate into practice and community with phase 4 clinical trials, clinical outcomes research, and population-level outcomes research.

Spinal cord stimulation

The therapeutic development of SCS is a noteworthy example of translational medicine and the bidirectional movement between pain research and clinical pain management.

Although randomized controlled studies support the efficacy of SCS, it did not evolve unidirectionally from bench to bedside. Despite Level 1 evidence for efficacy, the neurophysiologic and biochemical mechanisms of action were only partly elucidated and research moved from bedside to bench, where animal studies provided some evidence for spinal as well as supraspinal mechanisms of neuropathic pain relief of SCS.

Indeed, some of the research has progressed bench to bedside, including a SCS computer model of the electrical properties of the human spinal cord that revealed many basic neurophysiologic principles that were clinically validated later on.[19] Computer-based model studies also led to the development of electrode arrays with specific geometric properties, transverse tripolar stimulation, and development of a 5-column paddle lead. Mathematical modeling has highlighted the potential benefits of tight-electrode spacing in electrode contact design.[20]

Currently, the main question in clinical SCS is how to further improve the effectiveness of SCS as there is still a population of nonresponders. In this context, a return to preclinical studies is needed to elucidate which target pain neuron(s) are involved, as well as with what exact electrical stimulation this target neuron can be influenced to produce an optimal suppression of neuropathic pain.[19]

Analyzing current theories of pain mechanisms and their relevance to clinical pain

The first spinal cord stimulator was implanted in 1967, and the therapy has made steady progress in the last 50 years, with more significant advancements in the last decade. Technological advancements in lead design, refinements of anatomical targeting (including structures outside the cord itself), and novel waveforms such as burst SCS and high-frequency paresthesia-free SCS are likely to have contributed to this continuous improvement.[21] Recent developments include results presented from an RCT comparing Differential Target Multiplexed (DTM) to conventional SCS.[22] The hypothesized novel mechanism for DTM SCS is that glia are electrically responsive cells in the spinal cord and outnumber neurons by 12:1. Chronic pain occurs when neuro-glial interactions get out of sync. DTM SCS may impact the neuronal-glial interaction.[23] DTM SCS was superior to conventional SCS in reducing back pain, and results were sustained at 12 months.

Exploring new tools for both preclinical and clinical pain research

As noted above, outcomes from animal models used in preclinical research encounter translational difficulties. Experimental models tend to comprise genetically homogeneous groups of young, male rodents in restricted and unvarying environments, and pain-producing assays that may not closely mimic the natural condition of interest. In addition, typical experimental outcome measures using thresholds or latencies for withdrawal may not adequately reflect clinical pain phenomena pertinent to human patients. It has been suggested that naturally occurring disease in veterinary patients may provide more valid models for the study of painful diseases.

Many painful conditions in animals resemble those in people. Like humans, veterinary patients are genetically diverse, often live to old age, and enjoy a complex environment, often the same as their owners. There is increasing interest in the development and validation of outcome measures for detecting pain in veterinary patients; these include objective (e.g., locomotor activity monitoring, kinetic evaluation, quantitative sensory testing, and bioimaging) and subjective (e.g., pain scales and quality of life scales) measures. Veterinary subject diversity, pathophysiological similarities to humans, and diverse outcome measures could yield better generalizability of findings and improved translation potential, potentially benefiting both humans and animals. The study of natural animal models of pain could provide complementary information to that obtained from experimental models and address some of their limitations.[13] A veterinary model could help bridge the gap between bench and bedside.

The traditionally assumed translational approach for most research is applying what is learned in preclinical and clinical research to the clinic setting (bench to bedside). But with the shortcomings described above in having adequate animal models that address the psychosocial components of human pain, we are seeing utility in the reverse bedside-to-bench approach. For example, when successful neurostimulation techniques have been initially assessed in patients in the clinical setting they have then been translated to animal models for further investigation of their mechanisms of action. Many of the drug classes currently used to treat neuropathic pain, including antidepressants and antiepileptics, have been empirically supported for neuropathic pain, with subsequent evaluation of their mechanisms in animals.

Another bedside-to-bench translational approach in neuropathic pain consists of evaluating sensory phenotypes, used as surrogates of potential underlying mechanisms. This has led to a personalized therapeutic approach to neuropathic pain based on in-depth assessment of sensory profiles using questionnaires and quantitative sensory testing including dynamic quantitative sensory testing based on conditioned pain modulation. This approach is now endorsed by the European Medicines Agency. Furthermore "Patient Reported Outcome Measures" to improve management of pain are one of the major topics addressed by the large-scale ongoing IMI Pain Care consortium psychological phenotypes are expected to be as important as sensory phenotypes to predict treatment response in neuropathic pain and should probably be considered in future therapeutic studies. Most of the therapeutic advances achieved over the past decade in neuropathic pain still emanate from clinical observations, emphasizing the ongoing relevance of a bedside-to-bench translational approach.[24]

While the process of translational research can be complex and potentially unpredictable, its unique algorithmic approach aligns well with pain medicine. Because of its cumulative nature, it provides efficiency and effectiveness in the research process from moving a single or a combination of basic research findings into clinical practice with greater ease. It is yet another tool to help streamline the research process to meet the evolving demands of clinical pain medicine.

References

1. DeRenzo EG, Singer EA, Moss J. Special issues raised by evolving areas of clinical research. *Ethical Consid Preparing Clin Res Protoc*. 2020;271.
2. Hostiuc S, Moldoveanu A, Dascălu MI, Unnthorsson R, Jóhannesson ÓI, Marcus I. Translational research—the need of a new bioethics approach. *J Transl Med*. 2016;14(1). 1-0.

3. Schwartz J, Macomber C. So, you think you have an idea: A practical risk reduction-conceptual model for academic translational research. *Bioengineering*. 2017;4(2):29.
4. Cavalli E, Mammana S, Nicoletti F, Bramanti P, Mazzon E. The neuropathic pain: an overview of the current treatment and future therapeutic approaches.
5. Van Hecke O, Austin SK, Khan RA, Smith BH, Torrance N. Neuropathic pain in the general population: a systematic review of epidemiological studies. *Pain*. 2014;155 (4):654–662.
6. Cruccu G, Truini A. A review of neuropathic pain: from guidelines to clinical practice. *Pain Ther*. 2017;6(1):35–42.
7. Cavalli E, Mammana S, Nicoletti F, Bramanti P, Mazzon E. The neuropathic pain: an overview of the current treatment and future therapeutic approaches.
8. Deer TR, Pope JE, Hayek SM, et al. The Polyanalgesic Consensus Conference (PACC): recommendations for intrathecal drug delivery: guidance for improving safety and mitigating risks. *Neuromodulation*. 2017;20(2):155–176.
9. Deer TR, Mekhail N, Provenzano D, et al. The appropriate use of neurostimulation of the spinal cord and peripheral nervous system for the treatment of chronic pain and ischemic diseases: The Neuromodulation Appropriateness Consensus Committee. *Neuromodulation*. 2014;17(6):515–550.
10. Fishman MA, Antony A, Esposito M, Deer T, Levy R. The evolution of neuromodulation in the treatment of chronic pain: forward-looking perspectives. *Pain Med*. 2019;20 (suppl 1):S58–S68.
11. Krishnan Chakravarthy MD. Reframing the role of neuromodulation therapy in the chronic pain treatment paradigm. *Pain Phys*. 2018;21:507–513.
12. Available at www.harvardmedtech.com and www.harvardmedtech.com/outcomes/.
13. Klinck MP, Mogil JS, Moreau M, et al. Translational pain assessment: could natural animal models be the missing link? *Pain*. 2017;158(9):1633–1646.
14. Mao J. Translational pain research: bridging the gap between basic and clinical research. *Pain*. 2002;97(3):183–187.
15. Vadivelu N, Kai AM, Kodumudi V, Sramcik J, Kaye AD. The opioid crisis: a comprehensive overview. *Curr Pain Headache Rep*. 2018;22(3):16.
16. Staats PS, Yearwood T, Charapata SG, et al. Intrathecal ziconotide in the treatment of refractory pain in patients with cancer or AIDS: a randomized controlled trial. *JAMA*. 2004;291(1):63–70.
17. Wallace MS, Charapata SG, Fisher R, et al. Intrathecal ziconotide in the treatment of chronic nonmalignant pain: a randomized, double-blind, placebo-controlled clinical trial. *Neuromodulation*. 2006;9(2):75–86.
18. Rauck RL, Wallace MS, Leong MS, et al. Ziconotide 301 Study Group. A randomized, double-blind, placebo-controlled study of intrathecal ziconotide in adults with severe chronic pain. *J Pain Symptom Manag*. 2006;31(5):393–406.
19. Smits H, van Kleef M, Holsheimer J, Joosten EA. Experimental spinal cord stimulation and neuropathic pain: mechanism of action, technical aspects, and effectiveness. *Pain Pract*. 2013;13(2):154–168.
20. Hegarty D. Spinal cord stimulation: the clinical application of new technology. *Anesthesiol Res Pract*. 2012;2012.
21. Sivanesan E, Maher DP, Raja SN, Linderoth B, Guan Y. Supraspinal mechanisms of spinal cord stimulation for modulation of painfive decades of research and prospects for the future. *Anesthesiology*. 2019;130(4):651–665.
22. Fishman M., Cordner H, et al. DTM SCS RCT 12-month data results. Presented at a Medtronic webinar, jointly supported by the North American Neuromodulation Society

(NANS), World Institute of Pain (WIP), and the American Society for Pain and Neuroscience (ASPN); October 19, 2020. Webinar available on society websites.
23. https://www.medtronic.com/us-en/healthcare-professionals/therapies-procedures/neurological/spinal-cord-stimulation/education-training/dtm.html.
24. Attal N, Bouhassira D. Translational neuropathic pain research. *Pain*. 2010;160: S23–S28.

Ethical considerations in translational research

15

Nidhi Gupta
Department of Neuroanaesthesia, Indraprastha Apollo Hospitals, New Delhi, India

Introduction

Translational research, using first-in-man studies of experimental drugs/therapies, forms the basis of successful translation of basic science research from bench-to-bedside, and ultimately into clinical practice. Compared to standard clinical research, translational research involves a complex interplay of ethics, risk-benefit analysis, and autonomous decision-making on the part of researchers and study-subjects. While clinical medicine is solely based on providing optimal care to patients, the goal of translational research is to generate generalizable scientific data in humans after successful preclinical research in animal models. Nonetheless, while dealing with humans as research subjects, it is of utmost importance to adhere to certain ethical principles to ensure human safety, dignity, and quality of research.[1,2]

The main principles of translational research in human subjects rely on minimizing the risks to study-subjects while at the same time avoiding the missed clinical opportunities for important therapeutic advances because of their overprotection. Fundamentally, translational research requires certain ethical standards to guide researchers as well as approval and oversight of research proposals from independent committees, including a research ethics board, also known as Institutional Review Board (IRB) or Research Ethics Committee (REC).

This chapter focuses on a few of the many ethical issues that are relevant to translational research in general and for neurologically injured patients in particular.

Evolution of translational research ethics

Translation research involving human subjects has seen rapid growth since late 1800s due to unprecedented advancements in biomedical sciences and globalization of research. However, in the absence of regulatory law and guidelines, researchers used prisoners, children, mentally challenged, hospitalized, and terminally ill patients, and illiterate and poor people as their study-subjects, without their knowledge or informed consent. The global recognition and critic of these notorious scientific experiments post World-War II (WW-II) led to the evolution of ethics in

Perioperative Neuroscience. DOI: https://doi.org/10.1016/B978-0-323-91003-3.00004-0

translational research. Few such important trials which led to the creation of ethical codes of research conduct and regulatory bodies includes:

Nuremberg Code (1947)

The Nuremberg Code was the first international code of ethics that led to the introduction of concept of "informed consent" in the field of human subject's research. Throughout WW-II, numerous deadly and horrific human experiments were carried by the Nazi physicians and scientists on the victims of concentration camps on the pretext of scientific research. Post-WW-II (December 1946 to July 1947), the responsible team of physicians and medical staff were tried for crimes against humanity before a tribunal of three American judges in a trial known as "The Medical Case."[3] An American psychiatrist and neurologist Leo Alexander and another American physiologist Andrew Ivy, provided key expert testimony on the ethics of human experimentation, which served the basis for the documentation of Nuremberg Code. It has total 10 principles, the most notable being the absolute essentiality of voluntary consent of the human subjects, their sufficient knowledge and comprehension of the research intervention, ability to exercise free power of choice, an assessment of risks and benefits, assurances of competent investigators, need of preclinical animal studies, and termination of the experiments at any stage if this resulted in injury, disability, or death of the experimental subjects.[4]

Declaration of Helsinki (1964)

Working further on the basic ethical principles provided in the Nuremberg Code and the Declaration of Geneva, the World Medical Association (WMA) formally adopted the Declaration of Helsinki in 1964 during its 18th general meeting at Helsinki, Finland. It described the ethical principles for medical research involving human subjects, including research on identifiable human material and data. The declaration represents the international standard for the ethical treatment of human subjects and ensures that all research is conducted in a manner so that study-subject's safety is assured.[5] To date, the declaration has been revised multiple times, with the latest seventh version published during 64th WMA meeting in October 2013, Brazil. It provides for the compensation of the trial-related injuries, approval of use of placebos in the clinical trials, protection of vulnerable groups, and the posttrial provisions for the treatment that have shown benefits during the study.

Beecher's studies (1966)

Henry Knowles Beecher, professor of anesthesia research and chief anesthetist at Harvard, was one of the whistle blowers of the unethical human experimentation in Nazi concentration camps and a strong proponent for the rights of patients, human subjects, and scientific purity in the conduct of research.[6] He published a series of articles and presentations, describing in detail the unethical conduct of human research in uniformed soldiers, elderly patients, and mentally handicapped children.

These studies were carried out by well-known researchers from leading academic medical centers and published in high-impact scientific journals, primarily funded and sponsored by government and pharmaceutical companies. These studies brought out in open the abuse suffered by the most vulnerable people of society under the pretext of science and research.

The Belmont Report (1979)

Following Beecher's studies, the revelations of the Tuskegee Syphilis study[7] and Stanford Prison Experiment (1971),[8] further highlighted the unethical conduct of human-subject research, leading to the formation of the National Commission for the Protection of Human Subjects of Biomedical and Behavioral Research and the National Research Act (1974) by the United States Congress. The commission met in the Belmont Center of the Smithsonian Institution in April 1979 and issued its work as the "Belmont Report: Ethical Principles and Guidelines for the Protection of Human Subjects of Research."[9]

The report summarizes the most basic principles of biomedical research ethics, including autonomy and respect for people, beneficence (minimizing risks/harms and maximizing benefits from study participation), and justice (equitable distribution of benefits and risks/harms). The Belmont report was a landmark report, which formed the basis of the modern-day informed consent and research policies. It also added the concept of "proxy" or "surrogate consent" from a legally authorized representative within the provisions of the Declaration of Helsinki.

Common Rule (1991)

In 1991, the United States federal agencies formerly adopted the basic principles laid down in the Belmont report as the "Federal Policy for the Protection of Human Subjects," also known as the "Common Rule."[10] The Common Rule (45 CFR 46, Subpart A) governs research that is conducted on human subjects and is funded by one of 18 federal agencies. The rule requires the presence of voluntary informed consent by the human study-subjects, review of the research by IRBs, and institutional assurances of compliance with the regulations. As mandated by Common Rule, the IRBs can approve research protocols only when they ascertain reasonable risks to the human subjects in relation to anticipated benefits, if any. In its latest revision, the Pharmaceutical Research and Manufacturers of America Act of 2005 (H.R. 870) was introduced, to provide criminal penalties for concealing evidence of serious drug adverse events.[11]

International ethical guidelines

In 2002 the Council for International Organizations of Medical Sciences (CIOMS) along with the World Health Organization adopted the "International Ethical Guidelines for Biomedical Research," to unify the ethical codes for global clinical research. The document had 21 guidelines which primarily addressed the issues related to informed consent, ethical justification of the research, scientific validity, review and approval by IRB's, research in vulnerable human subjects and women participants, terms of confidentiality and compensation for research-related harms or injury. Later in 2011, the CIOMS Executive Committee decided to set up a Working Group to revise the CIOMS guidelines. The final and fourth version of the CIOMS guidelines were published in 2016, emphasizing the growing importance of translational research, acknowledging that exclusion of potentially vulnerable groups in many cases has resulted in a poor evidence base, clarifying what counts as fair research in low-resource settings, more emphasis on community engagement in research, and the increase of big data research.[12]

In October 2005, the United Nations Educational, Scientific and Cultural Organization (UNESCO) International Bioethics Committee adopted the "Universal Declaration on Bioethics and Human Rights" to lay down the global guidelines for bioethical values in relation to human dignity, rights, and freedom during scientific research.[13] The UNESCO declaration is considered significant as it is the first intergovernmental instrument on bioethics. It laid down new insistence on the relevance to the bioethics discipline for society, the community, humanity, all living-beings, and the biosphere. However, compared to other two international documents on bioethics (namely the WMA Declaration of Helsinki and the CIOMS guidelines on biomedical research), the UNESCO declaration has been criticized widely for its imprecise and noncommittal language.

Ethical considerations in translational research

There are certain critical ethical challenges that are crucial in translational research to support faster translation of preclinical research findings into innovative diagnostic, therapeutic, and preventive strategies (Table 15.1). These include:

High-quality preclinical research

Preclinical animal and laboratory research is an indispensable requirement before conducting first-in-man translational studies to minimize the unpredictable side effects or study-related risks to human subjects.[1] However, despite the quality and amount of preclinical data available, it is impossible to accurately predict the risks and possible clinical benefits of experimental drugs/therapies in humans.

In last few decades, advances in basic neurosciences have led to numerous successful preclinical studies in experimental models of traumatic brain injury (TBI) and stroke.[14,15] However, the translation of novel therapeutics from experimental models to clinical therapies has been extremely poor.[16,17] The varied reasons for the poor translation from bench-to-bedside in neuroscience has been attributed to unsatisfactory animal/experimental models, lack of alignment of discrete measures of the underlying injury pathology between the animal models and human subjects, unknown or little-known mechanism of action of drugs/therapies in humans, inherent differences at genetic level, heterogenous patient population, lack of suitable

Table 15.1 Basic principles of ethics in translational research.

• Respect for patient autonomy
• Voluntary informed consent
• Principle of beneficence
• Nonmaleficence ("first, do no harm")
• Justice
• Scientific validity
• Social value
• Equitable subject selection for research
• Clinical equipoise
• Protection of vulnerable subjects
• Independent ethical review by Institutional Review Boards/Research Ethics Committees
• Central review in multicentric studies
• Monitoring for data integrity and safety

outcome parameters, underpowered translational studies, and unavoidable in-between center differences for multisite studies.[18–19]

Apparently, for any research there is no way of knowing before starting a trial, whether the drug has benefits in humans or if it poses greater than minimal risks. Such uncertainties are further escalated in vulnerable populations including children, elderly, pregnant and lactating women. Nonetheless, despite the uncertainties, first-in-man trials must always be preceded by well-designed and pragmatically conducted laboratory studies to maximize human safety and successfully translate safe and effective therapies in humans.

Informed consent

The first and foremost ethical concern in translational research is the presence of a voluntary informed consent from all the participating human subjects.[20] An informed consent should include the fundamental ethical principles of respect for subject's autonomy, full disclosure of research and his/her ability to comprehend it. It should, therefore, provide meaningful information about the purpose and type of the research, the nature of the procedure or drug to be tested, risk and anticipated benefits, alternative treatment options available, information regarding data processing, and their rights to withdraw from the trial. The main goal of such a consent is to felicitate a true well-informed and voluntary decision-making by the subject, without any bias or coercion.

Ideally, an informed consent should be obtained in a language which is understood by the study-subjects, without the use of complex medical terminology. However, with the increasing complexity of translational research with large-scale genomic and other data sharing, the process of obtaining voluntary informed consents has become more and more complicated and challenging with time.[21] The subject's inability to comprehend further add challenges for people with low literacy and poor cognitive abilities, including the neurologically injured, mentally challenged, pediatric, and elderly participants. According to a literature review by

Sherlock et al., health care users' level of education, literacy, and language competency, combined with physicians' ability to effectively explain the medical procedure and the inherent risks and complications are important determinants of patients' capacity to provide fully informed consent.[22]

Addition of easy-to-read pamphlets and multimedia formats including audio and video demonstrations to the researcher's narrative have been found to significantly improve the study-subject's comprehension of the research protocol.[23] Furthermore, compared to the current paper-based consent, recently introduced electronic consents (e-consents) have the potential to improve the informed consent process in biomedical research.[24,25] The United States Food and Drug Administration (FDA) along with the Office for Human Research Protections, have issued recommendations regarding the process of obtaining e-consent including the validity of electronic signatures, verifying the identity of subjects or care-givers, sharing of information, and the review process by IRBs.[26] E-consent may facilitate long-term interaction with study-subjects for future follow-up studies and in obtaining consent from legal representatives or care-givers for vulnerable subjects and in emergency research.[25] As a pitfall, e-consent requires a certain level of literacy and multimedia support, which may not be possible in uneducated subjects and in low-income countries.[24] Furthermore, while dealing with e-consent, it is of utmost importance to protect subject's identity and data, especially with large-scale data sharing involved in genomic studies. Hence, though e-consent may supplement and expediate the existing paper-based consent process, it is still not standardized enough to replace it.

Therapeutic misconception

Therapeutic misconception is one of the most common and unresolved ethical issue in medical research.[1] It occurs when the study-subjects misinterpret the research protocol and believe that their participation in the research would be for their own therapeutic benefit, rather than to obtain generalizable scientific information. This ethical dilemma underscores the importance of a truly informed voluntary consent clearly stating the research goals, risks, possible benefits, and alternative treatment options to the potential research participants. Collaboration between researchers and treating physicians and continuous engagement of study participants in assessing risks and benefits of the research have been found useful in minimizing the therapeutic misbelief in early stage human-trials and translational genomic studies.[27]

Clinical "equipoise"

In biomedical research, the uncertainty regarding the relative effectiveness of the proposed experimental treatment compared to the existing treatment options, within the comparative arms of a randomized controlled trial (RCT), is called "equipoise." However, true equipoise rarely exists in clinical practice, leading to the concept of "clinical equipoise" by medical researchers, where there is a lack of consensus among the medical experts about the risk–benefit assessment of a treatment or drug.[28] Clinical equipoise is must for the ethical conduct of RCTs, as it nullifies the ethical question of causing risk to the subject by assigning him to a group that is known or believed to have an inferior treatment than the established standard of care.

The concept of "clinical equipoise" as a necessity for research has been a subject of great debate among the bioethicists.[29,30] According to Miller and Brody, clinical equipoise is neither necessary nor sufficient for ethical conduct of clinical trials and represents a fundamentally mistaken concept that promotes therapeutic misconception among study–subjects.[30] Furthermore, though it seems to avoid human subject exposure to known net risks associated with control group interventions, it allows research procedures with certain risks within the treatment group, adding to the ethical dilemma. Another major issue in maintaining clinical equipoise is that the flow of scientific information is dynamic. As new evidence becomes available during the course of trial, it may disrupt its clinical equipoise. Hence, the trial should be completed as early as possible to minimize the risk of losing its equipoise.

Incentive payments

Potential exploitation of healthy human volunteers by providing incentive payments is another common ethical concern in translational research.[31] The landmark Nuremberg Code states that the subject's participation in the trial should be voluntary and free from any coercion. Some researchers consider financial incentives given to healthy volunteers as coercive and unethical, while others claim it to be fair and rightful for any work performed on the part of subject.[31,32] In addition, financial payments may work as recruitment and retainment incentives for the adequate enrollment and follow-up of the trials.

Gelinas et al. recommends financial remuneration to economically vulnerable populations as ethical and justified, provided certain prerequisites are met with.[33] The study research protocol should obtain ethical approval from the concerned IRB in the absence of any consideration of compensation as a benefit and should have adequate safety precautions in place to obviate the risk of participants concealing important information or otherwise fabricating their medical history. Furthermore, the financial payments should be done as per regulatory requirements and should follow the local norms and practices.

Review by ethical boards

Worldwide, local IRBs/RECs are entrusted upon the responsibility of reviewing the research protocols to ensure that all the ethical and regulatory requirements are met with to protect human study-subjects from possible harm and exploitation. However, too strict adherence to ethical requirements by these regulatory bodies has been considered burdensome, time-consuming, and a hindrance to translational research.[34] Increasing costs, bureaucratic burden, lack of expertise, inflexibility, unclear goals, and limited evidence of effectiveness are other common problems attributed to IRBs. Evolving cell-based and genomic research, emergency research, multinational studies, and increasing complexity of oversight committees create additional challenges and responsibilities for IRBs, beyond those required by federal regulation.

The challenge now is to improvise the IRB process to keep in parallel with the emerging research methodologies. Many institutions opt for initial institutional pre-review of the protocol that precedes and complements the IRB review or for alternative models for IRB review such as shared or centralized review. Oversight committees may make recommendations on continuation or stopping a trial before completion, based on the evaluation of research data. Hence, an interim analysis of results must always be included in the study protocol to find about the futility of ongoing trial.

International collaboration

In parallel with rapidly growing basic sciences and genetic research, more and more study-centers need to be recruited to conduct the translational research with large sample sizes. However, ethical issues may arise in multicenter and multinational research because of the differences in the social and cultural norms, clinical practices, and the rules of regulatory bodies, especially for studies involving both developed and developing countries. Most multicentric RCTs require local IRB approval from each site, which may significantly delays the entire study process.[35] These trials thus require an effective collaboration between study-subjects, researchers, governments, and investors from different nations. Though the fundamental ethical principles of respect for persons and social justice remain paramount, the cultural and moral differences among people may have to be adjusted in the research protocol. To minimize inconsistencies in IRB decisions, central IRBs or standard evaluation sheet for IRB members may be created to evaluate the ethical appropriateness of proposed research protocols.[34,35]

Commercialization of translational research

One of the important responsibilities of research ethics boards is to prevent the possible commercialization of scientific research, a process called "translational medicine in reverse."[2] It involves possible manipulation of research goals with conversion of commercial products into scientific concepts without any evidence base.[36] Hence, it is of utmost importance for researchers to maintain a high level of integrity, ethics, and social responsibility when dealing with human subjects in translational research.

Ethical issues in vulnerable neurological population

Translational research in vulnerable human subjects mandate additional considerations. The vulnerable population for biomedical research includes mentally or cognitively impaired patients, children, old-age senile patients, pregnant women, fetuses, prisoners, uneducated, and economically disadvantaged people. Bioethicists have differing opinion regarding inclusion of vulnerable study population for first-in-man translational research, with majority of them advocating their exclusion from studies with more than minimal risks.[37] However, always excluding these patients would deprive them of their right to potential benefits of new drugs/therapies.[38] Children, elderly, and mentally incapacitated adults are often described as "therapeutic orphans" as they have been included in a limited number of translational studies that may yield useful results for them. This results in the widespread off-label use of drugs/therapies in this patient population on the basis of extrapolations from translational research in healthy human subjects.

Neurologically impaired patients come in the category of vulnerable study population, if they are unconscious or judged cognitively impaired because of the nature and/or extent of neurological injury. These patients may not be able to comprehend all the information disclosed in informed consent and are thus unable to make an autonomous decision about their participation.[39] Pape et al. have identified various areas of unresolved legal and ethical issues while conducting research in adults with severe TBI.[40] According to authors, the still lingering ethical issues in TBI translational research include

- determining the exact definition of vulnerability and voluntary informed consent in this patient population,
- determining decision-making capability of patients,
- the mistaken belief of therapeutic misconception by the grief-stricken care-givers/legal-surrogates,
- using care-givers as subjects, and
- neuroethics.

Emergency neurological research in patients with TBI, stroke, or intracranial hemorrhage is time-sensitive and requires their immediate enrollment in the ongoing trials. Failure to timely administer experimental drug/therapies has been implicated as one of the possible causes for the failure of translational research in TBI and stroke.[17,41] The research in neurologically injured patients must also account for the fluctuating nature of neurological recovery and associated memory loss. In these patients, the cognitive impairment or state of unconsciousness may last for an unpredictable course of time and the end point of recovery may be unknown. The subjects may be unconscious when enrolled into the protocol, but may regain their ability to comprehend and provide voluntary informed consent during recovery. Therefore the process for obtaining consent in neurologically injured patients must include methods for identifying and reevaluating subject's capacity to comprehend. To provide a valid consent, the patient, at the minimum, should be conscious with functional communication, use of an object, and/or a behavioral manifestation of a sense of self. However, at present there is no consensus among neuroscientists as to what state constitutes or does not constitutes consciousness.[42] For patients in early stages of a progressive neurodegenerative disease and with intact comprehension should be encouraged to provide voluntary consent for future interventions in the event of worsening or relapse of their illness.

Declaration of Helsinki allowed surrogate consent by a third party (next of kin or a legal representative) for subjects who are mentally and cognitively challenged. However, getting the surrogate consent is also a challenge in emergency settings for patients who are either unattended (such as victims of road traffic accidents) or have attendants who are unable to make a sudden thoughtful decision on emotional grounds. In 1996 exception from informed consent (EFIC) pathway was created by the USFDA to allow patient's enrollment without informed consent in some emergency trials. Since its creation, EFIC, also commonly referred to as waiver of informed consent, has led to considerable progress in emergency research on human subjects for certain life-threatening conditions, such as therapeutic hypothermia trials for TBI.[43–45]

The basic prerequisites for EFIC include emergency research on a life-threatening condition, subjects inability to provide voluntary informed consent, narrowness of timing of administration of experimental drug/therapy before surrogate consent can be obtained, nonavailability of alternative proven treatment options, preclinical research demonstrating low risk and potential benefits to the study–subjects, potential benefits must outweigh the risks, and possibility exists for obtaining surrogate consent from third party within the time-frame of therapeutic window. The concept of EFIC, however, adds to the ethical concerns for RCTs necessitating strict adherence to research protocol, especially for life-threatening conditions. In such scenarios, retrospective or prospective observational studies may provide an alternative practical approach for evaluating the effects of various treatment options.

The concept of "neuroethics" is new and still emerging in the field of TBI research.[39] Neuroethics has been defined as the "study of ethical, legal and social questions that emerge when scientific discoveries about the brain led to medical practices, legal interpretations and health and social policies"[46] Several sociobehavioral changes, including aggressiveness, irritability, and disinhibition, are common in individuals after sustaining TBI. Hence, neuroethical issues may arise during neurosciences research from the proposed interplay between brain and mind and its impact on individual's behavior such as logical reasoning, moral development, decision-making, and self-control.

Ethical concerns are also common in the genomic research involving neurocritically ill patients.[39] Translational genomic research in neurocritical care settings involves both forward and reverse genomic translational techniques in patients with ischemic and hemorrhagic stroke, TBI, subarachnoid hemorrhage, and status epilepticus.[47] Genomic studies undoubtedly improve our understanding of mechanisms of neurological injury and may thus help to identify the therapeutic targets and genetic risk-factors as predictors of patient prognosis. However, these studies generate a wealth of genomic data by examining the entire genome concurrently, and may prompt secondary data analysis. Ethical concerns in translational genomic research thus arise from the need to protect the patient's identity as well as to prevent large data sharing.[48] Sharing of highly sensitive genomic data with the patient may also be unethical as the primary purpose of these studies is not for genetic counseling and the data alone may be insufficient to make valid conclusions about its clinical impact.

Summary

The translational research from "bench-to-bedside" poses several critical ethical challenges which have direct implication in the safety and dignity of the human study–subjects. These include presence of voluntary informed consent, therapeutic misconception, and clinical equipoise. Over years, a number of legislations have been created as a result of human atrocities on the pretext of biomedical research including the Nuremberg Code, the Declaration of Helsinki, the Common Rule, and the Belmont Report. Each of these have served as a milestone in the creation of the notion of voluntary informed consent, the core of ethics in translational research. Regulatory bodies like IRBs also play a critical role in protecting human participants through prior and intermittent review of the ethical considerations of the research protocol.

Translational research in vulnerable neurologically injured patients add specific challenges in terms of emergency research and the inability of subjects to comprehend. Waiver of informed consent, surrogate consent, and e-consents from legal representatives are often used in the place of voluntary informed consent. The recently introduced translational genomic research further adds to the complexities of protecting patient's identity and large-scale data sharing. Finally, effective collaboration must be obtained between the researchers, scientists, review boards, funding agencies as well as the government bodies to ensure ethical conduct of scientifically rigorous translational research.

References

1. Petrini C. Ethical issues in translational research. *Perspect Biol Med.* 2010;53 (4):517–533. Available from: https://doi.org/10.1353/pbm.2010.0007. PMID: 21037406.
2. Sade RM. From laboratory to bedside: ethical, legal and social issues in translational research. *Am J Med Sci.* 2011;342(4):265–266. Available from: https://doi.org/10.1097/MAJ.0b013e318227deb1. PMID: 21952173; PMCID: PMC4517669.
3. Nuremberg Military Tribunals. 1949–1953. The Nuremberg code. In: Trials of War Criminals Before the Nuremberg Military Tribunals Under Control Council Law No. 10. Nuremberg, October 1946–April 1949. Washington, DC: GPO.
4. Schmidt U. *Justice at Nuremberg: Leo Alexander and the Nazi Doctors' Trial.* New York, NY: Palgrave Macmillan; 2004.
5. World Medical Association. World Medical Association Declaration of Helsinki: ethical principles for medical research involving human subjects. *JAMA.* 2013;310(20):2191–2194. Available from: https://doi.org/10.1001/jama.2013.281053. PMID: 24141714.
6. Kopp VJ. Henry Knowles Beecher and the development of informed consent in anesthesia research. *Anesthesiology.* 1999;90:1756–1765.
7. Brandt AM. Racism and research: the case of the Tuskegee syphilis study. *Hastings Cent Rep.* 1978;8(6):21–29. PMID: 721302.
8. Zimbardo PG. On the ethics of intervention in human psychological research: with special reference to the Stanford prison experiment. *Cognition.* 1973;2(2):243–256. Available from: https://doi.org/10.1016/0010-0277(72)90014-5. PMID: 11662069.

9. United States National Commission for the Protection of Human Subjects of Biological and Behavioral Research. The Belmont report: ethical principles and guidelines for the protection of human subjects of research. Department of Health, Education, and Welfare (DHEW) Publication No. (OS) 78–0012, Appendix I, DHEW Publication No. (OS) 78–0013, Appendix II, DHEW Publication No. (OS) 78-0014. Washington, DC: United States Government Printing Office; 1978.
10. HHS.gov. Office for Human Research Protections. Federal Policy for the Protection of Human Subjects ("Common Rule"). 45 CFR 46. 1981 Revised 1983, 1991, 2005.
11. Williams ED. Federal Protection for Human Research Subjects: An Analysis of the Common Rule and Its Interactions with FDA Regulations and the HIPAA Privacy Rule. 2005. <https://fas.org/sgp/crs/misc/RL32909.pdf>.
12. Council for International Organizations of Medical Sciences (CIOMS). International Ethical Guidelines for Health-related Research Involving Humans, Fourth Edition. 2016. <https://cioms.ch/wp-content/uploads/2017/01/WEB-CIOMS-EthicalGuidelines.pdf>.
13. UNESCO. Universal declaration on bioethics and human rights. Paris. 2006 <http://unesdoc.unesco.org/images/0014/001461/146180E.pdf>.
14. Stein DG, Sayeed I. Bridging the translational divide: emerging strategies in pharmacological approaches to traumatic brain injury. *Neuropharmacology*. 2019;145(Pt B): 131–132. Available from: https://doi.org/10.1016/j.neuropharm.2018.10.007. Epub 2018 Oct 9. PMID: 30308175.
15. Paul S, Candelario-Jalil E. Emerging neuroprotective strategies for the treatment of ischemic stroke: an overview of clinical and preclinical studies. *Exp Neurol*. 2021;335: 113518. Available from: https://doi.org/10.1016/j.expneurol.2020.113518. Epub 2020 Nov 2. PMID: 33144066; PMCID: PMC7869696.
16. O'Collins VE, Macleod MR, Donnan GA, Horky LL, van der Worp BH, Howells DW. 1,026 experimental treatments in acute stroke. *Ann Neurol*. 2006 Mar;59(3):467–477. Available from: https://doi.org/10.1002/ana.20741. PMID: 16453316.
17. Dietrich WD, Bramlett HM. Therapeutic hypothermia and targeted temperature management in traumatic brain injury: clinical challenges for successful translation. *Brain Res*. 2016;1640(Pt A):94–103. Available from: https://doi.org/10.1016/j.brainres.2015.12.034. Epub 2015 Dec 30. PMID: 26746342; PMCID: PMC4870128.
18. Cinelli P, Rauen K, Halvazishadeh S, et al. Translational research: what is the value of experimental studies in comparison with clinical studies to help understand clinical problems? *Eur J Trauma Emerg Surg*. 2018;44:645–647.
19. Sorby-Adams AJ, Vink R, Turner RJ. Large animal models of stroke and traumatic brain injury as translational tools. *Am J Physiol Regul Integr Comp Physiol*. 2018;315(2): R165–R190. Available from: https://doi.org/10.1152/ajpregu.00163.2017. Epub 2018 Mar 14. PMID: 29537289.
20. Lidz CW. Informed consent: a critical part of modern medical research. *Am J Med Sci*. 2011;342(4):273–275. Available from: https://doi.org/10.1097/MAJ.0b013e318227e0cc. PMID: 21804363.
21. Henderson GE. Is informed consent broken? *Am J Med Sci*. 2011;342(4):267–272. Available from: https://doi.org/10.1097/MAJ.0b013e31822a6c47. PMID: 21817873.
22. Sherlock A, Brownie S. Patients' recollection and understanding of informed consent: a literature review. *ANZ J Surg*. 2014;84(4):207–210. Available from: https://doi.org/10.1111/ans.12555. PMID: 24812707.
23. Perrenoud B, Velonaki VS, Bodenmann P, Ramelet AS. The effectiveness of health literacy interventions on the informed consent process of health care users: a systematic

review protocol. *JBI Database Syst Rev Implement Rep*. 2015;13(10):82–94. Available from: https://doi.org/10.11124/jbisrir-2015-2304. PMID: 26571285.

24. Chen C, Lee PI, Pain KJ, Delgado D, Cole CL, Campion Jr. TR. Replacing paper informed consent with electronic informed consent for research in academic medical centers: a scoping review. *AMIA Jt Summits Transl Sci Proc*. 2020;2020:80–88. PMID: 32477626; PMCID: PMC7233043.
25. De Sutter E, Zaçe D, Boccia S, et al. Implementation of electronic informed consent in biomedical research and stakeholders' perspectives: systematic review. *J Med Internet Res*. 2020;22(10):e19129. Available from: https://doi.org/10.2196/19129. PMID: 33030440; PMCID: PMC7582148.
26. United States Food & Drug Administration. Use of electronic informed consent in clinical investigations – questions and answers. Guidance for institutional review boards, investigators, and sponsors. 2016. <http://www.fda.gov/RegulatoryInformation/Guidances/default.htm>.
27. McCormick JB. How should a research ethicist combat false beliefs and therapeutic misconception risk in biomedical research? *AMA J Ethics*. 2018;20(11):E1100–E1106. Available from: https://doi.org/10.1001/amajethics.2018.1100. PMID: 30499440.
28. Miller FG. Clinical equipoise and risk-benefit assessment. *Clin Trials*. 2012;9(5): 621–627. Available from: https://doi.org/10.1177/1740774512450952. Epub 2012 Jul 9. PMID: 22777654.
29. Djulbegovic B. Uncertainty and equipoise: at interplay between epistemology, decision making and ethics. *Am J Med Sci*. 2011;342(4):282–289. Available from: https://doi.org/10.1097/MAJ.0b013e318227e0b8. PMID: 21817885; PMCID: PMC3183244.
30. Miller FG, Brody H. A critique of clinical equipoise. Therapeutic misconception in the ethics of clinical trials. *Hastings Cent Rep*. 2003;33:19–28.
31. Halpern SD. Financial incentives for research participation: empirical questions, available answers and the burden of further proof. *Am J Med Sci*. 2011;342(4):290–293. Available from: https://doi.org/10.1097/MAJ.0b013e3182297925. PMID: 21817878.
32. Bernstein SL, Feldman J. Incentives to participate in clinical trials: practical and ethical considerations. *Am J Emerg Med*. 2015;33(9):1197–1200. Available from: https://doi.org/10.1016/j.ajem.2015.05.020. Epub 2015 May 29. PMID: 26095131; PMCID: PMC4562811.
33. Gelinas L, White SA, Bierer BE. Economic vulnerability and payment for research participation. *Clin Trials*. 2020;17(3):264–272. Available from: https://doi.org/10.1177/1740774520905596. Epub 2020 Feb 17. PMID: 32063065.
34. Grady C. Institutional review boards: purpose and challenges. *Chest*. 2015;148(5): 1148–1155. Available from: https://doi.org/10.1378/chest.15-0706. PMID: 26042632; PMCID: PMC4631034.
35. Ekmekci PE, Guner MD. Inconsistencies in institutional review board decisions: a proposal to regulate the decision-making process. *Bratisl Lek Listy*. 2019;120(2):95–101. Available from: https://doi.org/10.4149/BLL_2019_015. PMID: 30793611.
36. Munsie M, Hyun I. A question of ethics: selling autologous stem cell therapies flaunts professional standards. *Stem Cell Res*. 2014;13(3 Pt B):647–653. Available from: https://doi.org/10.1016/j.scr.2014.04.014. Epub 2014 Apr 29. PMID: 24857592.
37. Bracken-Roche D, Bell E, Macdonald ME, Racine E. The concept of "vulnerability" in research ethics: an in-depth analysis of policies and guidelines. *Health Res Policy Syst*. 2017;15(1):8. Available from: https://doi.org/10.1186/s12961-016-0164-6. Erratum in: Health Res Policy Syst. 2017 Apr 3;15(1):29. PMID: 28173859; PMCID: PMC5297186.

38. Dakić T. No research for the decisionally-impaired mentally ill: a view from Montenegro. *BMC Med Ethics*. 2020;21(1):47. Available from: https://doi.org/10.1186/s12910-020-00489-z. PMID: 32517748; PMCID: PMC7285582.
39. Hicks R. Ethical and regulatory considerations in the design of traumatic brain injury clinical studies. *Handb Clin Neurol*. 2015;128:743–759. Available from: https://doi.org/10.1016/B978-0-444-63521-1.00046-7. PMID: 25701918.
40. Pape TL, Jaffe NO, Savage T, Collins E, Warden D. Unresolved legal and ethical issues in research of adults with severe traumatic brain injury: analysis of an ongoing protocol. *J Rehabil Res Dev*. 2004;41(2):155–174. Available from: https://doi.org/10.1682/jrrd.2004.02.0155. PMID: 15558370.
41. Cheng YD, Al-Khoury L, Zivin JA. Neuroprotection for ischemic stroke: two decades of success and failure. *NeuroRx*. 2004;1(1):36–45. Available from: https://doi.org/10.1602/neurorx.1.1.36. PMID: 15717006; PMCID: PMC534911.
42. Septien S, Rubin MA. Disorders of consciousness: ethical issues of diagnosis, treatment, and prognostication. *Semin Neurol*. 2018;38(5):548–554. Available from: https://doi.org/10.1055/s-0038-1667384. Epub 2018 Oct 15. PMID: 30321893.
43. Klein L, Moore J, Biros M. A 20-year review: the use of exception from informed consent and waiver of informed consent in emergency research. *Acad Emerg Med*. 2018; 25(10):1169–1177. Available from: https://doi.org/10.1111/acem.13438. Epub May 17, 2018. PMID: 29679443.
44. Clifton GL, Miller ER, Choi SC, et al. Lack of effect of induction of hypothermia after acute brain injury. *N Engl J Med*. 2001;344(8):556–563. Available from: https://doi.org/10.1056/NEJM200102223440803. PMID: 11207351.
45. Clifton GL, Valadka A, Zygun D, et al. Very early hypothermia induction in patients with severe brain injury (the National Acute Brain Injury Study: hypothermia II): a randomised trial. *Lancet Neurol*. 2011;10(2):131–139. Available from: https://doi.org/10.1016/S1474-4422(10)70300-8. Epub 2010 Dec 17. PMID: 21169065; PMCID: PMC3628679.
46. Safire W. Visions for a new field of "neuroethics.". In: Marcus SJ, ed. *Neuroethics: Mapping the Field*. New York: The Dana Press;; 2002:3–9.
47. Myserlis P, Radmanesh F, Anderson CD. Translational genomics in neurocritical care: a review. *Neurotherapeutics*. 2020;17(2):563–580. Available from: https://doi.org/10.1007/s13311-020-00838-1. PMID: 32080794; PMCID: PMC7223188.
48. Oliver JM, Slashinski MJ, Wang T, Kelly PA, Hilsenbeck SG, McGuire AL. Balancing the risks and benefits of genomic data sharing: genome research participants' perspectives. *Public Health Genom*. 2012;15(2):106–114. Available from: https://doi.org/10.1159/000334718. Epub 2011 Dec 30. PMID: 22213783; PMCID: PMC3318928.

Index

Note: Page numbers followed by "*f*" and "*t*" refer to figures and tables, respectively.

9780323910033